Agriculture and Environment in India

Agriculture and Environment in India

By

Surya Bhushan

Associate Professor of Economics,
Development Management Institute (DMI), Patna, Bihar

New Century Publications
New Delhi, India

New Century Publications
4800/24, Bharat Ram Road,
Ansari Road, Daryaganj,
New Delhi - 110 002 (India)

Tel.: 011-2324 7798, 4358 7398, 6539 6605
Fax: 011-4101 7798
E-mail: indiatax@vsnl.com • info@newcenturypublications.com
www.newcenturypublications.com

Editorial office:
LG–7, Aakarshan Bhawan,
4754-57/23, Ansari Road, Daryaganj,
New Delhi – 110 002

Tel.: 011-4356 0919

First Published: **2016**

ISBN: **978-81-7708-420-7**

Published by New Century Publications and printed at Salasar Imaging Systems, New Delhi.

Designs: Patch Creative Unit, New Delhi.

PRINTED IN INDIA

About the Book

The most important natural resource of India is land which is the basis for all agricultural and allied activities. While the population grows, the land surface is fixed and only a certain proportion of it is available for cultivation. In the context of increasing human and animal population, the non-expandable land resources have to accommodate the competing demands for production of food, fodder, fibre, fuel, minerals, urbanisation and other non-agricultural uses.

Land degradation, occurring due to the natural and human induced causes, is one of the priority concerns in India. The varying degrees and types of degradation stem mainly from unsustainable use and inappropriate land management practices.

Wastelands, deserts, drought-prone and flood-prone areas have been subjected to severe forms of degradation. The capacity of these lands is limited due to environmental factors. Pressures of human and livestock population have further compromised them.

Direct consequences of agricultural development on environment arise from intensive farming activities, which contribute to soil erosion, land salination and loss of nutrients. Modern agricultural practices in the country have led to over-exploitation of land and water resources and excessive use of fertilizers and pesticides. Shifting cultivation has also been a major factor responsible for land degradation in hilly areas.

It is essential to control soil erosion in order to attain and maintain food security, sustainable forestry and agricultural and rural development.

About the Author

Dr. Surya Bhushan is currently Associate Professor of Economics at the Development Management Institute (DMI), Patna, Bihar. Previously, he worked with the marketing analytics division of Accenture Management Consultancy. He did his M.A., M.Phil. and Ph.D. in Economics from the Centre for Economic Studies and Planning (CESP), Jawaharlal Nehru University (JNU), New Delhi.

He has authored and presented a number of papers in reputed refereed journals on agricultural productivity measurement. His current research interest includes agricultural development, productivity measurement and livelihood resilience.

Contents

About the Book v
About the Author vi
Preface .... xi-xiii

1. Role of Agriculture in Economic Development **1-9**
1.1 Effect of Agriculture on Economic Growth
1.2 Arguments for Investment in Agriculture
1.3 Theodore W. Schultz on Agriculture
1.4 Agriculture in Developing Countries
1.5 Agricultural Productivity

2. Indian Agriculture: An Overview **10-31**
2.1 Brief History
2.2 Importance of Agriculture in the Indian Economy
2.3 Subsistence Nature of Farming and Small-sized Holdings
2.3.1 Reasons for Small Holdings
2.3.2 Disadvantages of Small Holdings
2.3.3 Farm Size and Productivity
2.4 Cropping Pattern in Indian Agriculture
2.4.1 Classification of Agricultural Crops
2.5 Low Productivity of Indian Agriculture
2.5.1 Causes of Low Productivity
2.6 Declining Share of Agriculture in Gross Domestic Product (GDP)
2.7 Eleventh Five Year Plan (2007-12) on Agriculture
2.8 Oilseeds/Edible Oils: Areas of Concern
2.9 Post-harvest Losses

3. Phases of Agricultural Development in India **32-38**

3.1 First Phase: Pre-Green Revolution Period (1950-51 to 1967-68)

3.2 Second Phase: Green Revolution Period (1968-69 to 1980-81)

3.3 Third Phase: Post-Green Revolution (1981-82 to 1990-91)

3.4 Fourth Phase: Post-reform Period (1991-92 to 2000-01)

3.5 Fifth Phase: Period of Recovery (2001-02 to 2010-11)

4. Productivity and Efficiency of Indian Agriculture **39-50**

4.1 Labour-intensive Agriculture

4.2 Agriculture Productivity and Rural Incomes

4.3 Demand for Workers

4.4 Agricultural Inputs

4.5 Structural Transformation

5. Review of Studies on Agricultural Productivity **51-68**

6. Environmental Sustainability of Indian Agriculture **69-80**

6.1 Environment and Development: The Conflict

6.2 Environmental Pollution

6.3 Sustainable Development: The Balancing Act

6.4 Agriculture-Environment Interface

6.5 Climate Change and Agriculture

6.6 National Mission for Sustainable Agriculture (NMSA)

6.7 Prime Minister's Council on Climate Change

6.8 Conclusions

7. Degradation of Soil Resource Base **81-107**

7.1 Problem of Land Degradation

7.2 Land Pollution and Land Degradation: Causes
7.2.1 Intensive Agricultural Practices
7.2.2 Unsanitary Disposal of Solid Wastes
7.2.3 Pressure of Human and Cattle Population
7.2.4 Community Lands
7.2.5 Mining Activities
7.2.6 Industrial Effluents
7.2.7 Over Extraction from Forests
7.2.8 Other Causes
7.3 Consequences of Land Degradation
7.4 State-wise Soil Fertility Status
7.5 Soil Erosion and Productivity Loss
7.6 Environmental Implications of the Green Revolution
7.7 Pest Problems
7.8 Strategies to Check Land Degradation
7.9 Programmes/Schemes for Treatment of Degraded Lands
7.9.1 Soil and Land Use Survey of India (SLUSI)
7.9.2 Soil Conservation Training Centre, Damodar Valley Corporation (DVC), Hazaribagh, Jharkhand
7.9.3 Soil Conservation in the Catchments of River Valley Projects and Flood Prone Rivers
7.9.4 Reclamation and Development of Alkali and Acid Soils (RADAS)
7.9.5 Watershed Development Project in Shifting Cultivation Areas
7.9.6 National Mission for Sustainable Agriculture (NMSA)
7.10 Fly Ash Utilization
7.11 Mining and Land Use Management

8. Degradation of Water Resources **108-125**
8.1 Importance of Water
8.2 Utilization of Water
8.2.1 Agricultural Demand for Water
8.2.2 Industrial Demand for Water
8.2.3 Domestic Demand for Water
8.3 Ground Water Irrigation
8.4 Ground Water Depletion
8.5 Farm-level Effects of Irrigation-induced Salinity
8.6 Water Use Efficiency
8.7 Strategy for Efficient Management of Water Resources

9. Summary and Conclusions **126-132**

Appendix: Edited Extracts on Agriculture from Twelfth Five Year Plan (2012-17) **133-144**

Bibliography **145-176**

Index **177-182**

Preface

The most important natural resource of India is land which is the base for agricultural production. After China, India is the most populous country in the world accounting for 16.7 percent of world population. It is the seventh largest country in the world occupying 2.4 percent of total world area. With a total land area of 3,287,263 square kilometres, it measures 3,214 kilometres from North to South and 2,993 kilometres from East to West. It has a land frontier of 15,200 kilometres and its sea coast runs to the length of 7,517 kilometres.

While population grows, the land surface is fixed and only a certain proportion of it is available for cultivation. In the context of increasing human and animal population, the non-expandable land resources have to accommodate the competing demands for production of food, fodder, fibre, fuel, minerals, urbanisation, non-agricultural land use etc.

In the past, increase in net sown area came about by converting forest area into cultivable area. There is no such scope at present or in the near future. The only alternative available is to reclaim the waste land and the land suffering from salinity, alkalinity and water logging. In reclaiming available wasteland, private sector initiative and expertise should be tapped wherever necessary.

Land degradation occurring due to the natural and human induced causes, like wind erosion and water logging, is one of the priority concerns in India. The varying degrees and types of degradation stem mainly from unsustainable use and inappropriate land management practices.

Land, a non-renewable resource, is central to all primary production systems. Over the years, India's landmass has suffered from different types of degradations. Floods and droughts are recurrent phenomena in India, severely constraining agriculture. The problem of land degradation gets exacerbated due to large scale land erosion in which fertile top soil is lost.

Intensive agricultural practices that rely heavily on water, chemical fertilisers and pesticides have caused water-logging and salinity in many parts of the country. The expansion of the irrigation system without adequate steps for treatment of the catchment areas has exacerbated this. The quest for increased agricultural productivity has led to intensive cultivation of marginal lands causing their degradation.

Pollution of land results also from the unsanitary disposal of solid wastes. In India open dumping of such wastes of municipal and industrial origin on low lying land is a common phenomenon. This serves as a breeding ground for pests and disease carrying vectors.

In addition to the wastelands, other areas such as deserts, drought-prone, flood-prone and tribal areas have been subjected to severe forms of degradation. The capacity of these lands is limited due to environmental factors. Pressures of human and livestock population have further compromised them.

Land degradation is considered to be an unavoidable by-product of mining and has reached alarming proportions, mainly due to over-exploitation and mismanagement of natural resources. Mining activity often leads to environmental problems like land degradation, particularly in open cast mining and land subsidence in underground mining. Open cast mining in areas with forest cover causes deforestation.

Soil pollution from heavy metals due to improper disposal of industrial effluents, along with the excessive use of pesticides and mismanagement of domestic and municipal wastes, is becoming a major concern. Though no reliable estimates are available to depict the exact extent and degree of this type of land degradation, it is believed that the problem is extensive and its effects are significant. Some commercial fertilizers also contain appreciable quantities of heavy metals, which have undesirable effects on the environment. The indiscriminate use of agrochemicals, such as fertilizers and pesticides, is often responsible for land degradation.

Land degradation has a direct bearing on the productivity of soil, its vulnerability to rainfall variations, scarcity of drinking water, fodder and fuel wood. Given the inter-linkages of crop production, livestock economy and environment, land degradation has a major impact on the livelihoods of the people, especially in rural areas.

Direct consequences of agricultural development on the environment arise from intensive farming activities, which contribute to soil erosion, land salination and loss of nutrients. The introduction of *Green Revolution* in the country has been accompanied by over exploitation of land and water resources and excessive usage of fertilizers and pesticides. Shifting cultivation has also been a major factor responsible for land degradation in hilly areas. Leaching due to extensive use of pesticides and fertilizers is a major source of contamination of water bodies.

The extent of agricultural intensification and extensification is characterized by an increase in cropping and irrigation intensity and the unbalanced use of chemical fertilizers, pesticides and insecticides. It has also led to land degradation, over exploitation of underground water resources and increased use of chemical fertilizers, leading to eutrophication and water pollution in some regions.

Enhanced intensification and extensification also leads to salination, alkalization and water logging in irrigated areas, along with eutrophication of water bodies and ill health of oceans, leading to loss of biodiversity. For achieving and maintaining food security and sustainable forestry, controlling of land/soil erosion is extremely vital. It is essential to control soil erosion in order to attain and maintain food security, sustainable forestry and agricultural and rural development.

Surya Bhushan

Patna
October 2015

1

Role of Agriculture in Economic Development

Concerns about the ability of agriculture to provide sustainability for the needs of human population are by no means new. Generations of development economists have stressed improvement in agricultural productivity as an essential part of the development strategy. [1] For example, T.R. Malthus (1798) argued that, because of the law of diminishing marginal productivity, agriculture would be limited in its ability to feed an ever-increasing population. Later writers, such as David Ricardo (1817), argued that, with technical or scientific progress and sufficient capital investment in agriculture, the Malthusian problem would not be a real issue.

Initially, Malthus argued that population would only decline through 'natural checks' such as famine and pestilence, although in his later writings (1830), he amended this to include falling birth rates as a result of 'moral restraint'. Malthus also argued that growing populations would expand into more marginal land, and returns to labour would inevitably decline. He noted that any improvements in production techniques, which occurred more by chance than anything else, would induce further population growth, raise labour supply and food demand but because of the diminishing conversion efficiency of labour (and land) into food, such improvements effectively cancel themselves out. [2] Following Malthus, some have argued that population growth rates are such that it will impossible to maintain adequate food supplies, and environmental decline will inevitably result.

1.1 Effect of Agriculture on Economic Growth

The potential contribution of agriculture to economic

growth has been an ongoing subject of much controversy among development economists. Much of the early work on this issue coincided with the debate on the role of agriculture in promoting economic development in low-income nations in the aftermath of extended periods of colonial rule (Lewis, 1954, Fei and Ranis, 1961; Jorgenson, 1961; Johnston and Mellor, 1961; Schultz, 1964). Nurkse (1953, p. 52) argued that "everyone knows that the spectacular Industrial Revolution would not have been possible without the agricultural revolution that preceded it", and Rostow (1960, p. 8) stated that "revolutionary changes in agricultural productivity are an essential condition for successful take-off". [3]

Much of these investigations were qualitative in nature and they emphasized the potential impact of the inter-sectoral linkages between agricultural and industrial manufacturing sectors. After a lull in research on this subject, the recent flurry of theoretical and empirical studies on the subject indicates that the debate has increased in intensity (Echevarria, 1997; Humphries and Knowles, 1998; Gemmell, Lloyd, and Mathew, 2000; Kogel and Prskawetz, 2001; Gollin, Parente and Rogerson, 2002; Olsson and Hibbs, 2005; Tiffin and Irz, 2006). Research on this issue is crucial because it helps inform domestic and international policy makers regarding how scarce resources are allocated to agricultural research and infrastructure.

Nevertheless, recent empirical studies have yielded mixed and sometimes conflicting evidence and there remains a lack of consensus on the effect of agriculture on economic growth. Masterfully synthesizing the main lessons from nearly five decades of such analysis, Staatz and Eicher (1998, p. 31, chapter 2) explain, "By the end of the 1990s, development thinking had come nearly full circle. In the 1950s and 1960s, many development economists analyzed how the agricultural and non-agricultural sectors interacted during the process of economic growth, using simple two-sector models. This abstract theorizing was sharply criticized by dependency theorists, among others, who argued that such work abstracted from the institutional and structural barriers to broad-based growth in most low-income countries.

During the 1970s and 1980s, the focus of research shifted to developing a more detailed theoretical and empirical understanding of the rural economy. But the emphasis on structural adjustment in the 1980s forced re-examination of agriculture's relationship to the macro economy. By the late 1990s, economists were again focusing on how the rural economy was linked to the broader world market, but they demonstrated a renewed recognition of how important institutions are in determining a country's pattern of growth and the distribution of the benefits of that growth.

1.2 Arguments for Investment in Agriculture

In the ensuing decade, the profession finally got agriculture back on the broader development agenda. A key breakthrough was the publication by the World Bank (2007) of the World Development Report (WDR) 2008, 'Agriculture for Development'. Coordinated and drafted by Derek Byerlee and Alain de Janvry, this WDR was the first in a quarter century to focus specifically on agriculture. Its publication late in 2007, just as the world food crisis was heating up, looked prescient. Still, none of the major donor agencies have figured out how to gear up quickly to support more spending on agricultural development, partly because there remains deep uncertainty over what to do and how to do it.

In an influential paper, Johnston and Mellor in 1961, which marks a change in development thought, lay out the following role of agriculture in the development process of an economy.

First, they argue that agriculture must generate a food supply sufficient to satisfy a growing demand. If this growth is not fulfilled, then economic growth can be compromised. In an intuitive exposition, the annual growth in food demand (D) is given by $D = p+n.g$, where p is the annual population growth rate, g is the annual per capita income growth rate, and n is the income elasticity of demand for food. The growth in D is of special importance in developing countries like India given that the values for p and n tend to be relatively high.

Second, agriculture is producing not only for the internal consumption but also for the export markets. In this fashion, this sector can contribute to the generation of foreign exchange, which can be devoted to importing capital goods that cannot be produced domestically.

Third, it can transfer human resources to the city. Initially, this transfer is relatively easy if the farm labour supply is abundant and its marginal productivity is low. This notion is a major implication of the Lewis' (1954) surplus labour model. As the country develops, the transfer of workers must be accompanied by an increase in labour productivity through farm mechanization.

Fourth, it can contribute to capital formation throughout the economy. Again, an important insight of the Lewis (1954) model is the explanation of why the rate of capital accumulation is critical in determining the pace at which the industrial sector can generate employment. A poor country that is trying to enhance its economic growth finds a dire need to augment its capital in order to develop its infrastructure. The predominance of the agriculture sector clearly suggests its importance in generating the resources necessary to undertake such investments.

Finally, Johnston and Mellor (1961) contend that agriculture can be a significant stimulus to the industrial sector by acting as a major consumer of locally produced manufactured products.

1.3 Theodore W. Schultz on Agriculture

In the voluminous literature generated thereafter on the role of agriculture in development, it is important to single out the Schultz's (1964) monograph entitled *Transforming Traditional Agriculture*. Theodore W. Schultz argued and advanced the hypothesis that farm families in developing countries were "efficient but poor" and thus that "there are comparatively few significant inefficiencies in the allocation of the factors of production in traditional agriculture" (Schultz, 1964; pp. 37-38). The argument set forth by Schultz is that peasant farmers behave as rational economic agents, in the neo-classical sense, and evaluate the costs and benefits associated with different production techniques, and thus these farmers allocate their resources efficiently.

According to this argument, it is not possible to obtain noticeable gains in output by simply reallocating existing resources. In other words, he contends that traditional peasant farmers, despite their low level of output per capita, are allocatively efficient. This hypothesis is an enduring view in the literature on development economics. It has led policy makers to believe that the improvement could not be achieved since the farmers adhere to their traditional technologies. Consequently, this has resulted in policies, emphasizing investments in generating new and more productive technologies. Thus, the corollary of this hypothesis is that gains in productivity in a traditional agricultural setting can only be obtained through technological change stemming from research and extension services.

1.4 Agriculture in Developing Countries

However, the farmers in developing countries, like India, with low literacy rates, facing poor extension services, lack of credit and capital and inadequate physical infrastructure have great difficulties in understanding and adopting new technologies. The introduction of new technologies is not a one point phenomenon so long as there is advancement and innovation in new technologies. This will put the recipients of the technology in disequilibria due to the introduction of technologies in a continuous pace. Empirical results suggest that farmers in developing countries fail to exploit fully the potential of technology making inefficient decisions. Allocative efficiency is usually measured by comparing the marginal value of product of an input to its price as a ratio capital.

Bhatia (1967) constructed a measure of agricultural efficiency, suitable for areas where livestock did not constitute an integral part of the agricultural production. The study focused on the agricultural practices in the province of Uttar Pradesh, India, using data over the period 1960-61. Agricultural efficiency was stated to be a function of various factors that included the physical (climate and soil), socio-economic (size of holding and type of farming), and technical (crop rotation, irrigation and mechanization). The author used an indexing method, classifying the efficiency index into high, medium, low and very low. This was done by first estimating yield efficiency of each crop and expressing acre-yields as a percentage of average acre-yields. Then agricultural efficiency was measured by the weighted average of all the yield efficiencies using the share of cropland devoted to each crop as the weight. A total of 11 crops, representing 86.4 percent of cropland, were selected for this study—these included rice, wheat, barley, jowar (sorghum), bajra (millet), maize, chick peas, arhar (a pulse), peas, sugar cane, and groundnut (peanut). The results showed that the yields of rice, wheat, barley, bajra, grams, and sugar canes displayed wide disparity of efficiency within the province. In the case of rice, barley, bajra and grams, high density areas did not generally show high acre-yield, unlike wheat and sugar cane. Finally, there was a wide disparity in spatial efficiency with more efficient regions recording up to thirty percent higher than the average yield. The low and medium efficient regions generally skirted around the continuous belts of the very low and highly efficient regions.

1.5 Agricultural Productivity

Studies by Shapiro (1983) and Junankar (1980) rejected the hypothesis that farmers were allocatively efficient in using their resources. Technical inefficiency, measured indirectly, is more frequently used in the literature than direct estimates of technical efficiency. There are no apparent empirical regularities in the direct technical inefficiency estimates reviewed from various studies. Rather than directly estimating technical efficiency, some studies measured the effect of non-conventional inputs, such as education, on technical efficiency. Sahota (1968) estimated an average capital value of 0.8 for bullock labour and 1.05 for fixed capital for Uttar Pradesh and Bombay, whereas Barnum and Squire (1978) estimated a capital value of 2.7 for variable inputs (presumably fertilizer) for the India as a whole, and concluding that the farmers were allocatively efficient.

Using the primary data collected from Palanpur village in UP, Bliss and Stern (1982) estimated a capital value of 3.5 for fertilizer inputs, while Feder et al. (1987), for Haryana, found an increase in wheat productivity by 1 percent per year due to schooling. The training and visit extension system also increased productivity by 9 percent. In another study, Sidhu and Baanante (1979), in case of Punjab, found that an additional year of farmer education makes an estimated contribution of 1.7 percent in wheat production, and the coefficient of education on agricultural productivity was 3.6 percent. Chaudhri (1976) also found a high coefficient of education of 11.5 percent on agricultural productivity for Punjab, Haryana, and Uttar Pradesh. Kalirajan and Shand (1989) estimated for Indian paddy farm and found farming experience, education, and access to credit and extension as statistically significant factors. [4] In general, most of the efficiency studies concluded that lack of technical knowledge and education are primary sources of technical and allocative inefficiencies, which implies that increasing the efficiency of input-use by improving farmer knowledge and skills provides a potential for productivity growth in the medium- to long-term. Thus, the "poor but efficient hypothesis" may have overlooked important potential for increasing agricultural productivity.

Currently, policy makers have started to believe that an important source for growth of the agricultural sector is efficiency gain through greater technical and allocative efficiency by producers, in response to better information and education. Technical change and efficiency improvements are very important factors of productivity growth especially in developing agricultural economies, where resources are meagre and opportunities for developing and adopting better technologies have lately started dwindling (Ali and Chaudhury, 1990; Ruttan, 2002). Furthermore, current projections are that income and population growth will double the demand for agricultural products over next 50 years. Therefore, substantial gains in productivity will be needed to keep up with the increase in demand (Ruttan, 2002). Vision 2020 document of Planning Commission of Government of India also acknowledged the vital role of agriculture in the national development.

Endnotes

1. Agriculture continues to impact on rural poverty. Datt and Ravaillion (2009) demonstrated that agricultural growth remains an important determinant of rural poverty reduction. This conclusion is, further, reinforced by the same regression analysis of Himanshu et al. (2011) that showed that higher yields are associated with declining rural poverty, suggesting the impact of agricultural productivity growth on poverty remains high.
2. Malthus pre-empted Boserup, as in his model any extra income, due to Boserupian intensification, would merely set population growth off again. Easter Boserup (1965) argued that increasing population pressure provides the primary stimulus for innovation and intensification. Key responses to population pressure include increased cropping intensities (with shorter and less frequent fallows) and the introduction of land saving techniques. She has defined agricultural intensification as 'the gradual change towards patterns of land use which make it possible to crop a given area of land more frequently than before'. Boserup's argument focuses on a number of stages in the frequency of cropping—each stage has different cultivation techniques and the model implies a progression from less to more intensive cultivation.

   Thus, increased population pressure results in a shift from forest fallow- to bush- then grass fallow- annual and finally multicropping. At heart of the model is the notion of technological change induced or impelled by a 'critical' population density. She concedes that very high population growth rates are unlikely to fit this model. Examining Boserup's thesis, Pingali and Binswanger (1988) show that there is a 'remarkable degree of substitutability of capital and labour for land, so that in the long run returns to agricultural labour appears to decline quite slowly as population density increases'. They suggest that farmer generated technical change is capable of sustaining slow and steady population increases with modest increases in agricultural output, but may not be capable of supporting rapidly rising populations. At this stage, large scale technical changes need to take place. They note that while farmer based innovations and the intensification of agricultural systems are constrained by agro-climatic conditions, a pattern can be observed in the degree of investment in the land.

3. This view had gained the status of axioms in development economics as evident from some important works, Gillis *et al.* (1983), Hayami and Ruttan (1985), Herrick and Kindleberger (1983), Tummer (1988) and Todaro (1989), Ray (2000). Timmer claims that this view "has not been challenged".
4. Battesse (1992), and Bravo-Ureta and Pinheiro (1993) have reviewed the literature relating to the technological change and technical efficiency in the agriculture of developing world.

2

Indian Agriculture: An Overview

2.1 Brief History

Agriculture in India has a long history, dating back to 9000 BC as a result of early cultivation of plants and domestication of crops and animals. Double monsoons led to two harvests being reaped in one year. Until British colonial rule, the Indian economy had been known for centuries for its self-contained village communities, consisting of agriculturists, cottage industrialists, village craftsmen, artisans, unskilled workers and village officials. These communities played a major role in not only meeting the needs of the village economy but producing and exporting various products to foreign countries. During those times, agriculture was a way of living and the farmers produced merely for self-consumption. Food crops like wheat and rice were the most important ones. Since plants and animals were considered essential to their survival, people started worshipping and respecting them.

The middle ages saw irrigation channels reach a new level of sophistication in India and Indian crops affecting the economies of other regions of the world under Islamic patronage (Iqtidar and Shaffer). Land and water management systems were developed to provide uniform growth (Kumar). However, unlike several 'white' or 'European' colonies across the Atlantic or in the Pacific, the Indian subcontinent was not for domicile but primarily for the benefit of the colonial masters. Prosperity of the native settlers was not an objective of colonial rule but a possible externality of the colonial design of development.

The British Empire supported improving the productivity of land with irrigation projects, reservoir dams and river diversions in several regions, particularly in the Indo-Gangetic plain and parts of the coastal peninsula. [1] Revenue settlement and other institutions governing lease, user rights and property rights over land very substantially influenced the choice of regions for irrigation development in addition to their topography. Hence, major irrigation was devised in some parts rather than others. Impressive engineering feats for the control of water for irrigation and other use were achieved during that era. [2] The cultivation technology however remained virtually the same, which continued into the early decades of post-colonial India. [3]

2.2 Importance of Agriculture in the Indian Economy

At the time of Independence in 1947, Indian agriculture was traditional and stagnant in every respect. It was characterized by feudal land relations, primitive technology, and the resultant low productivity per hectare. The first task of Indian Government in the immediate post-Independence period was, therefore, to initiate growth process in agriculture on modern lines. Modernization of agriculture was required both in terms of technological and institutional changes. Abolition of intermediaries in agriculture, like *zamindars* and *jagirdars*, was accomplished soon after Independence.

The largest portion of the natural resources of India consists of land and by far the largest number of its inhabitants is engaged in agriculture. Therefore, in any scheme of economic development of the country, agriculture holds a position of basic importance. Although Indian agriculture is way back compared to the levels in developed countries, some notable developments have occurred over the post-Independence period. Large areas which suffered from repeated failures of rainfall have received irrigation, new crops have come to occupy a significant position in the country's production and trade, agricultural and the industrial economies in the country now exert a powerful influence on one another, problems of rural indebtedness and the exploitative practices of the village moneylenders are much less, and finally there is already in the countryside an awakening and a desire for improved standard of living.

Agricultural sector occupies a key position in the Indian economy. It provides employment to about 60 percent of the working population of India. Around 14 percent of India's national income originates from the agricultural sector. It is imperative that the problems of farmers are addressed with a sense of urgency. Agriculture being a State subject, the bulk of public investment in agriculture takes place at the level of States and the Central Government supports the States as a catalyst.

The First Five Year Plan (1951-56) accorded the highest priority to the agricultural sector to tide over the difficult food problem created by the partition of the country. Agriculture has occupied an important place in every successive plan. Two major components of agricultural development strategy have been: (a) subsidies on inputs and (b) minimum support price for output.

Agricultural products like cereals (mainly rice), tea, coffee, cashew, spices, tobacco and leather are important items of India's exports and hence foreign exchange earnings. Agriculture is also the source of raw material for agro-based industries including textiles, cigarettes, jute, sugar, paper, processed foodstuffs and *vanaspati*.

Moreover, agricultural sector provides market for capital goods (tractors, pump sets and other agricultural machinery), inputs (fertilisers, insecticides), and light consumer goods. The nation has invested huge resources for the development of agriculture under various plans.

The development of the agricultural sector depends, to a large extent, on such core industries as power, petroleum, fertilisers and machine tools. Thus, there is inter-dependence between agriculture and industry.

Agriculture being a state subject, the primary responsibility for increasing agriculture production, enhancing productivity, and exploring the vast untapped potential of the sector rests with the State Governments. However, in order to supplement the efforts of the State Governments, a number of Centrally-sponsored and Central Sector schemes are being implemented for enhancing agricultural production and productivity in the country and increasing the income of the farming community.

2.3 Subsistence Nature of Farming and Small-sized Holdings

As against the commercialised nature of agricultural operations in the developed countries, farming in India still remains a source of livelihood for most farmers in India. The main reason for this is the low per capita availability of land. Per capita availability of land (or land-man ratio) declined from 0.91 hectare in 1951 to 0.48 hectare in 1981 and to 0.27 hectare in the year 2011 (Table 2.1).

Table 2.1: Declining Per Capita Availability of Land (Land-man Ratio) in India

Year	Total geographical area (million hectares)	Total population (million)	Col.2/Col.3 Per capita availability of land (hectare)
1	2	3	4
1951	328.73	361	0.91
1961	328.73	439	0.75
1971	328.73	548	0.60
1981	328.73	685	0.48
1991	328.73	844	0.39
2001	328.73	1006	0.33
2011	328.73	1210	0.27

The average size of farms has become smaller over the years and the trend continues. An important reason for this trend is the growth of population which has adversely affected the per capita availability of land. The pressure of population along with some social and economic factors has decreased the size of agricultural holdings in India. Various surveys conducted by official agencies point out this fact (Table 2.2). Moreover, traditional agricultural practices still prevail in large parts of the country. Use of modern agricultural technology has remained confined to few pockets of affluent farmers in select States.

Table 2.2: Trends in the Average Size of Holdings in India

(hectares)

NSS Eighth Round (1953-54)	3.0
NSS Seventeenth Round (1961-62)	2.6
Agricultural Census (1975-76)	2.0
Government of India (1990-91)	1.6
Government of India (2000-01)	1.3
Agricultural Census (2005-06)	1.23
Agricultural Census (2010-11)	1.16

Sources: (i) Government of India, Planning Commission, *Sixth Five Year Plan* (1980-85), Annexure 9.1, p. 139 and (ii) Government of India, Ministry of Agriculture, *Agricultural Statistics at a Glance*, 2012.

2.3.1 Reasons for Small Holdings: There are various reasons for sub-division and fragmentation of holdings in India.

1. The laws of inheritance in India give to all children an equal share in the property of their parents. After the enactment of the Hindu Succession Act, 1956, daughters have also become, along with sons, entitled to an equal share in the property of their parents. Consequently, land is being sub-divided with every passing generation.
2. As already noted, the population of India has increased rapidly after Independence. It increased from 36.1 crore in 1951 to 43.9 crore in 1961, to 54.8 crore in 1971 and to 68.5 crore in 1981. The census of 1991 enumerated the population at 84.4 crore which increased to 100.6 crore in 2001 and was enumerated at 121 crore in the 2011 Census. However, the area of agricultural land has remained almost the same. Moreover, due to lack of employment opportunities in the non-agricultural sector, most of the

additional labour force has to fall back on agriculture for earning a living.

3. Decline of joint family system has also contributed to the break up of agricultural holdings. Legal heirs prefer to have their individual shares in family property instead of holding it as joint property of the family. The rise of individualism in modern societies has hastened the decline of joint family system.
4. Rural indebtedness has also contributed to the sub-division and fragmentation of holdings. As is well-known, most farmers in India are indebted to moneylenders and landlords. In the event of non-payment of their debt, they are forced to sell a portion of their land to the creditor.
5. It is a common practice in India to lease out land to tenants on crop-sharing basis. Quite often the landlords lease out their land to a number of tenants to avoid trouble and also to circumvent the land laws. Due to these practices, land gets sub-divided and fragmented.

2.3.2 Disadvantages of Small Holdings: Sub-division of holdings leads to loss of cultivable land due to boundaries. Too many boundaries cause disputes, often leading to litigation. Instead of investing their precious resources in improving the productivity of land, farmers waste their limited funds in settling claims and counter claims. This tendency has created tensions in rural areas.

Fragmented holdings create difficulties in the management of land resources. When a farmer has to shift from one piece of land to another, it involves wastage of manpower and extra expenses. Similarly, small-sized farms restrict opportunities for the use of modern science and technology. Proper arrangements for irrigation cannot be made if farms belonging to a farmer are scattered here and there. Modern agricultural machines cannot be successfully operated on tiny farms. Absence or limited use of modern agricultural practices keeps farm productivity low.

2.3.3 Farm Size and Productivity: What is the ideal size of a holding? The ideal size of a holding (called economic holding or optimum holding) is one which allows the most efficient use of land, labour and capital with a given technique of production. It is difficult to prescribe a uniform size of economic holding for all crops and for all regions. In fact, the size of economic holding depends upon a host of factors such as nature of the crop, fertility of soil and technique of production.

There is a great debate among agricultural economists in India regarding the relationship between size of farms and productivity/profitability. Productivity simply means output per hectare of land while farm-profitability relates to the value of output over all costs of cultivation. Some experts have argued that as farm size increases, productivity also increases. However, some researchers hold the opposite view and maintain that productivity declines as farm size increases. Although the debate is inconclusive and the picture confusing, it has been reasonably well established by statistical studies that farm size and productivity are inversely related. The myth that small farms are inefficient has been exploded. The opponents of land ceiling laws always criticised the break up of large farms on efficiency ground. Since the efficiency of small farms has nearly been confirmed, the case for land ceiling on equity ground has become stronger.

2.4 Cropping Pattern in Indian Agriculture

Cropping pattern refers to the distribution of cultivated land among different crops grown in a country. Cropping pattern reveals the nature of agricultural operations, e.g. the importance of food crops vis-à-vis cash crops. Cropping pattern is influenced by a host of factors which can be broadly classified into two categories: (a) physical factors and (b) economic factors.

Among the physical factors, the important ones are soil conditions, extent of rainfall and type of climate. The economic factors include relative prices of agricultural commodities, size of the farms, availability of inputs, demand conditions, system of land holding and government policy regarding exports, imports, taxes and subsidies.

There are two main agricultural seasons in India: (a) *kharif* under which crops are planted at the onset of the Southwest monsoon in June-July and harvested in September-October and (b) *rabi* under which crops are planted usually between October and December and harvested between March and May. Major *kharif* crops are rice, jowar, bajra, maize, cotton, sugarcane, soyabean and groundnut. Major *rabi* crops are wheat, barley, gram, linseed, rapeseed and mustard.

India is a large country with diverse climatic, soil and terrain conditions. A wide variety of crops are grown in different parts of the country.

2.4.1 Classification of Agricultural Crops: Due to its rich biological diversity, various types of crops are grown in India. The important among these may be classified as under:

A. Foodgrains: These include rice, wheat, coarse cereals (like maize, gram, *bajra, jowar)*, and pulses. These crops constitute the backbone of India's food economy. In India, nearly 75 percent of the cultivated area is occupied by food crops including rice, wheat, millets and pulses. Paddy (rice) is the most important food crop in India followed by wheat. There has been no significant change in the cropping pattern during the post-Independence period. The present cropping pattern is unbalanced as it is heavily tilted in favour of paddy and wheat while area under oilseeds and pulses is less than required.

B. Oilseeds: There are ten major oilseeds produced in India, namely groundnut, castorseed, sesame, nigerseed, soyabean, sunflower, rapeseed, mustard, linseed and safflower. Of these, groundnut, rapeseed, mustard and soyabean occupy commanding position in terms of total production.

C. Fibre Crops: Cotton and jute are the chief crops under this category.

D. Plantation Crops: These include sugarcane, tea, coffee, tobacco, rubber, cocoa, cardamom, cashew and coconut. Some of these crops constitute important items of India's exports. India is the largest producer, processor, consumer and exporter of cashew in the world. Coconut is an important tree crop grown extensively in coastal areas providing livelihood to about 10 million people in cultivation, processing and related activities.

E. Horticultural Crops: India, with its wide variability of climate and soil, produces a large range of horticultural crops such as fruits, vegetables, potato, tropical tuber crops, ornamental crops, medicinal and aromatic plants, spices (like black pepper, cardamom, ginger, garlic, turmeric and chillies). The National Horticulture Mission was launched on April 1, 2005. The Mission ensures end-to-end approach having backward and forward linkages covering research, production, post-harvest management, processing and marketing.

Indian agriculture has indeed diversified from foodgrains to a variety of crops but more needs to be done in this regard. In his 2005-06 budget speech, the Finance Minister announced the preparation of a roadmap by the Ministry of Agriculture to focus on fruits, vegetables, flowers, dairy, poultry, fisheries, pulses and oilseeds.

2.5 Low Productivity of Indian Agriculture

Indian agriculture was backward and stagnant at the time of Independence. Ever since the launching of the First Five Year Plan (1951-56), agricultural sector has received the prime attention of the Government in the overall strategy of economic development. As a result, farm productivity has increased over the years and the country has achieved high degree of self-sufficiency in terms of foodgrains and raw material for agro-based industries (Tables 2.3 and 2.4). Although per hectare yield of major crops has increased over the last four decades yet it is far below the international levels. This scenario calls for an examination of the causes of low productivity of agriculture in India.

2.5.1 Causes of Low Productivity: Statistical evidence

suggests that farm productivity in India is low as compared to other countries of the world. The low productivity of the agricultural sector is attributable to a host of technological and institutional factors. Some of the important of these factors are described below.

Table 2.3: Trends in the Productivity of Indian Agriculture: Selected Years

(yield per hectare/kg.)

Crop	1970-71	1980-81	1990-91	2000-01	2014-15
Rice	1,123	1,336	1,740	1,901	2,390
Wheat	1,307	1,630	2,281	2,708	2,872
Coarse cereals	665	695	900	1,027	1,729
Pulses	524	473	578	544	744

Source: Reserve Bank of India, *Handbook of Statistics on Indian Economy, 2014-15*, Table 22 (excerpted).

Table 1.5: Index of Agricultural Production in India

(Base: Triennium ending 1981-82 for data till 2000-01)
(*Base: Triennium ending 2007-08=100, for data 2010-11 onwards)

1950-51	1960-61	1970-71	1980-81	1990-91	2000-01	*2010-11	*2012-13	*2013-14
46.2	68.8	85.9	102.1	148.4	165.7	121.0	124.2	129.2

Source: Government of India, *Economic Survey*, 2014-15, Statistical Appendix, p. A21, Table 1.9 (excerpted).

A. Technological Factors: These include the following.

- **Lack of Irrigation Facilities:** Despite best efforts during the last five decades, only one-third of cultivated land has the facility of artificial irrigation. In other words, most of the farmers depend on uncertain rainfall for irrigation of their fields. Rainfall is uncertain in terms of quantity, place and time. Agriculture is still a gamble in monsoon for millions of farmers in this country.
- **Limited Use of Fertilisers:** Adequate use of fertilisers is crucial for improving the farm productivity. Unfortunately, the use of fertilisers by Indian farmers is much below the level achieved in developed countries.

- **Limited Use of High Yield Variety of Seeds:** The HYV programme was started in 1966. The percentage of HYV area to total area differs from crop to crop. In the case of wheat it is quite high. However, the coverage of area under HYV for other crops is relatively low.
- **Inadequate Plant Protection Schemes:** Another reason for low productivity is the damage caused to crops by insects, pests, and plant diseases. The use of pesticides and insecticides is not widespread, particularly among the small and marginal farmers. Central Integrated Pest Management Centres have been set up in various States to monitor crop pests, and to organise training in pest management for extension workers and farmers.
- **Lack of Farm Mechanisation:** Most Indian farmers still use age-old farming techniques like wooden plough, cattle power, and sickle. The use of modern farm machinery is limited to affluent farmers in selected regions of the country. There is a great deal of controversy regarding mechanisation of agriculture in view of widespread unemployment and disguised unemployment in rural areas. Therefore, the policy regarding mechanisation of agriculture has to be cautious and selective.
- **Floods and Soil Erosion:** Floods are a common occurrence in one or the other part of the country every year. The consequent soil erosion and waterlogging dislocates production cycle, resulting in loss of agricultural output.

B. Institutional Factors: Use of modern methods of production is conditioned by the institutional framework of a production sector. If the institutional set up is exploitative, it will discourage the adoption of efficient technology. Unfortunately, the institutional setting of the farm sector in India is not congenial for the use of modern agricultural technology.

- **Feudal Land Relations:** Although *zamindari* system was abolished soon after Independence, it continues to exist in a different garb. Land reform laws have proved ineffective and therefore big landlords still rule the roost in rural areas. They lease out their land to tenants and charge high rents. The various tenancy malpractices hardly provide any incentive to tenant farmers to improve land and productivity.

- **Small-sized Holdings:** As a result of population growth, there is tremendous pressure on available cultivable land. This has led to overcrowding and hence sub-division and fragmentation of holdings. Small-sized and scattered holdings imply wastage of time and labour, difficulties in the use of modern techniques, and quarrels and litigation among farmers. The existence of these problems is also responsible for poor farm yield.
- **Rural Indebtedness:** Poverty of the Indian farmer forces him to borrow money for his existence. The institution of money lending is the greatest curse for the poor and illiterate rural folk of India. Although the importance of moneylenders has declined over the last 55 years, they are still very dominant in providing credit to the farmers. The exploitative practices of the moneylenders are no secret. Their malpractices include charging of high rate of interest and adopting unfair means in the maintenance of accounts.
- **Marketing Difficulties:** Marketing is the last ring in the production chain. If the marketing system is inefficient and denies the farmers a reasonable return for their efforts, they will be discouraged to produce more. Unfortunately, the present marketing system of agricultural produce is not favourable to the interests of the farmers. The preponderance of intermediaries and the malpractices of commission agents deny remunerative prices to the farmers.

Thus, low productivity in Indian agriculture is the result of a number of technological and institutional factors which influence each other. The causes of low agricultural productivity explained above suggest their own remedies.

2.6 Declining Share of Agriculture in Gross Domestic Product (GDP)

Over the years, however, the share of agriculture in GDP has fallen significantly, and as a result the aggregate GDP has become relatively less sensitive to fluctuations in agricultural performance. This factor is of course well recognised, but the role of agriculture, and particularly agricultural incomes, in influencing GDP growth indirectly through the demand side is less well understood.

Highlighting the changes in the sectoral composition of Gross Domestic Product (GDP) over the years, the *Economic Survey*, 2011-12 observed, "The structure of the economy has also undergone significant changes over time. Between 1950-51 and 1980-81, the industrial sector registered a higher growth rate than the services sector. The converse has been the case since then. This resulted in the share of the industry sector in GDP increasing by around 9 percentage points from 16.6 percent to 25.9 percent during the period from 1950-51 to 1980-81. The share of the services sector increased from 30.3 percent in 1950-51 to 38 percent in 1980-81. It started growing rapidly thereafter and this phenomenon became more pronounced in the 1990s. Consequently, since 1980-81, the share of the industry sector has remained in the range of 26 to 28 percent of GDP, while the entire decline in share of agriculture has been balanced by an increase in share of the services sector. Thus, the resilience of the economy to shocks owe to the services sector which has the largest share and most consistent growth performance". [4]

The Document further noted, "Notwithstanding the declining trend in agriculture's share in GDP, the importance of the sector to the economy is best understood with reference to its share in employment and in terms of its criticality for macroeconomic stability. While the former was well known, the latter became manifest with rising growth in incomes since the mid-2000s.

Hence, growth in agriculture and allied sectors remains an

important objective and a 'necessary condition' for inclusive growth". [5] The changes in relative shares of various sectors in GDP are shown in the Table 2.5.

Table 2.5: Changes in the Sectoral Composition of GDP

Year	Agriculture	Industry	Services
1950-51	53.1	16.6	30.3
1960-61	48.7	20.5	30.8
1970-71	42.3	24.0	33.8
1980-81	36.1	25.9	38.0
1990-91	29.6	27.7	42.7
2000-01	22.3	27.3	50.4
2012-13	13.9	27.3	58.8
2013-14*	13.9	26.1	59.9

* Provisional.

Source: Government of India, *Economic Survey*, 2013-14, p. 6.

2.7 Eleventh Five Year Plan (2007-12) on Agriculture

Expressing concern on agricultural crisis and outlining strategy to regain agricultural dynamism, the Approach Paper on Eleventh Five Year Plan released by the Government in November 2006 observed, "One of the major challenges of the 11th Plan will be to reverse the deceleration in agricultural growth from 3.2 percent observed between 1980 and 1996-97 to a trend average of around 2 percent subsequently. This deceleration is the root cause of the problem of rural distress that has surfaced in many parts of the country and reached crisis levels in some. Low farm incomes due to inadequate productivity growth have often combined with low prices of output and with lack of credit at reasonable rates to push many farmers into crippling debt. Even otherwise, uncertainties seem to have increased (regarding prices, quality of inputs and also weather and pests) which, coupled with unavailability of proper extension and risk insurance have led farmers to despair. This has also led to widespread distress migration, a rise in the number of female headed households in rural areas and a general increase in women's work burden and vulnerability. In 2004-05, women accounted for 34 percent of principal and 89 percent of subsidiary workers in agriculture, higher than in any previous round of the National Sample Survey.

The crisis of agriculture is not a purely distributional one arising out of the special problems of small and marginal farmers and landless labour. In fact, agricultural deceleration is affecting farms of all sizes.

To reverse this trend, corrective policies must not only focus on the small and marginal farmers who continue to deserve special attention, but also on middle and large farmers who suffer from productivity stagnation arising from a variety of constraints.

It is vital to increase agricultural incomes as this sector still employs nearly 60 percent of our labour force. A measure of self-sufficiency is also critical for ensuring food security. A second green revolution is urgently needed to raise the growth rate of agricultural GDP to around 4 percent. This is not an easy task since actual growth of agricultural GDP, including forestry and fishing, is likely to be below 2 percent for the Tenth Plan period. The challenge posed, therefore, is to at least double the rate of agricultural growth and to do so by recognising the demographic realities, particularly the increasing role of women". [6]

Outlining its strategy to raise agricultural output, the Plan document stated, "The Eleventh Plan strategy to raise agricultural output will be based on the following elements:

- Double the rate of growth of irrigated area.
- Improve water management, rain water harvesting and watershed development.
- Reclaim degraded land and focus on soil quality.
- Bridge the knowledge gap through effective extension.
- Diversify into high value outputs, fruits, vegetables, flowers, herbs and spices, medicinal plants, bamboo, bio-diesel etc., but with adequate measures to ensure food

security.

- Promote animal husbandry and fishery.
- Provide easy access to credit at affordable rates.
- Improve the incentive structure and functioning of markets.
- Refocus on land reforms issues. [7]

2.8 Oilseeds/Edible Oils: Areas of Concern

The success the country achieved in raising production of foodgrains, sugar, fruits, vegetables and various plantation crops perhaps lulled the country into believing that the shortfall in oilseeds and pulses could likewise be successfully addressed. The Technology Mission on Oilseeds (1986) has had some initial success but accelerated annual GDP growth of 6 percent in later years as also continued expansion in population (1.9 percent) caused demand for oils to rise (now 100-110 lakh tonnes annually), which the domestic output (60-70 lakh tonnes) is unable to meet. Edible oil is an essential commodity, more so when a large majority of Indians are vegetarians and animal fats cannot be a substitute. Import of over 50 lakh tonnes annually meets almost half the country's demand currently.

Raising domestic production of three principal oil crops—groundnut, mustard and soyabean—has serious limitations. First major limitation is the lack of any genetic advance in technology in evolving high productivity seeds. Second limitation is the compulsion of raising oilseeds on unirrigated soils (80 percent of area) and the third limitation is low productivity per hectare which can never match rice/wheat productivity levels even if grown on irrigated soil and finally the high risk in losing production due to extreme sensitivity to adverse weather (frost/hailstorm) especially in case of mustard crop.

Unlike rice, wheat, sugar, bajra, jowar and cotton, farmers do not and cannot raise oilseed as a principal crop because of extreme risk factor. It is a supplementary cash crop and a proverbial gamble amidst weather uncertainty. Price is an important issue. Diversion of area from foodgrains to oilseeds is widely talked about and the key to this diversion is thought to lie in raising MSP of oilseeds to provide oilseed growers a comparative price advantage relative to rice and wheat. Unfortunately, this is unlikely to help as long as yield rates in oilseeds are so low relative to rice and wheat. Doubling or even tripling the MSP of oilseeds will not match their gross return per hectare vis-à-vis foodgrains.

Given the current low productivity levels in three major oilseeds—groundnut, mustard and soyabean, and the limitations to raising productivity levels—the country's dependence on imports to meet almost half of its requirements is unavoidable. Doubling of international prices in 2002 caused little or no reduction in Indian imports which is evidence of inelasticity of demand at current levels of GDP and population growth in the country. High import duties too had little effect on imported volumes. Improvements in yield rates and favourable pricing policy together with a high WTO compatible tariff wall can at best make only a marginal impact. Dependence on large volume of imports is bound to continue.

2.9 Post-harvest Losses

A comprehensive nation-wide quantitative assessment of harvest and post-harvest losses for 46 agricultural produces was carried out to estimate the extent of harvest and post-harvest losses. This assessment included five cereals, four pulses, six oilseeds, eight fruits, eight vegetables, sugarcane, four seed spices, three plantation crops, milk, meat, inland and marine fish, poultry meat and eggs. To assess these losses, data were collected through integrated stratified multistage survey design from 106 randomly selected districts of the country representing all targeted agricultural produces. Two blocks from each district and five villages from each block were randomly selected as second and third stage sampling units. Finally, ten farmers from each village were taken for the assessment. This assessment covered 14 out of 15 agro-climatic zones (Island region agro-climatic zone was not included). The data were collected through enquiry as well as actual observations. Estimates obtained through both procedures were combined by assigning appropriate weighing factors before arriving at final estimate of losses for a commodity. The operations considered for assessment of losses were harvesting, collection, threshing, grading/sorting, winnowing/cleaning, drying, packaging, transportation, and storage depending upon the commodity. The post-harvest losses in cereals, pulses, oilseeds, fruits, vegetables, spices, condiments (plantation crops) are shown in Table 2.6.

Table 2.6: Post-harvest Losses of Select Agricultural Outputs in India

S.No.	Produce	Loss in Percent
1.	Cereals	3.9-6.0
2.	Pulses	4.3-6.1
3.	Oilseeds	2.2-10.1
4.	Fruits	5.8-18.0
5.	Vegetables	6.8-12.5
6.	Spices	3.6-6.8
7.	Condiments (Plantation Crops)	0.6-6.9

Source: Government of India, Ministry of Agriculture, Department of Agricultural Research and Education, *Annual Report*, 2010-11, p. 57.

To sum up, agricultural policy followed since Independence can be broadly classified into three phases. Phase one, immediately after independence, witnessed several institutional changes through land reforms and initiation of major irrigation projects. The second phase, since the mid-1960s, was aimed at increasing agricultural productivity to attain self-sufficiency in foodgrains through technological revolution and public support to credit, marketing, extension etc. This was followed by the current phase, since the late 1980s, aimed at making agriculture more market-oriented.

Agriculture sector is vital for the food and nutritional security of the nation. The sector remains the principal source of livelihood for more than 58 percent of the population though its contribution to GDP has declined to 14.2 percent due to high growth experienced in industries and services sectors. Compared to other countries, India faces a greater challenge, since with only 2.3 percent share in world's total land area, it has to ensure food security of its population which is about 17.5 percent of world population. This leads to excessive pressure on land and fragmentation of land holdings.

Unfortunately, Indian economy is witnessing the slowing of agriculture at a time when the rest of the economy is growing at unprecedented rates. The sector, as a whole, has showed poor performance in the recent years. Agricultural growth rate has declined and foodgrains production has virtually stagnated. Signs of agrarian distress are visible in some years. The spate of suicides by farmers in many areas is the most disconcerting manifestation of this distress. Marginal and small farmers have borne the brunt of the adverse circumstances in agriculture.

As regards the performance of agricultural sector in recent Plans, the Approach Paper to Twelfth Five Year Plan (2012-17) observed, "A weakness in the economic performance thus far is that growth in the farm sector (agriculture and allied activities), though better than in the Tenth Plan, remains short of the 4.0 percent Plan target. The farm sector has grown at an average rate of around 3.2 percent during the first four years of the Eleventh Plan and assuming conditions remain favourable in 2011, the average farm sector growth in the Eleventh Plan period may be a little over 3.0 percent. This is a marked improvement from the average growth of about 2.0 percent during the Tenth Plan period. Still, with half of our population dependent on agriculture and allied activities, we need faster farm sector growth to benefit poor farmers, many of whom are women. The below target growth in this sector is one of the reasons for increase in food prices over the last two years". [8]

While agriculture no longer dominates the Indian economy in terms of national output, it is still the most important in terms of employment. In other words, declining share of income has not been accompanied by an equivalent decline in employment. It is a classic case of lop-sided structural transformation. Accelerating the rate of growth of agricultural production is seen as central to a more inclusive growth.

Although agriculture is still the backbone of the Indian economy, at least in terms of employment, the impulses of reforms have been relatively less in scope and depth in this sector. It is noteworthy that reforms in agricultural sector were introduced only towards the end of the 1990s. Nevertheless, a series of policy initiatives have been undertaken in this sector as well. These include: (a) partial decontrol of fertiliser prices, (b) removal of bottlenecks in agricultural marketing, (c) relaxation of restrictions under the Essential Commodities Act, 1955, (d) introduction of forward trading in important commercial crops, (e) replacement of the Revamped Public Distribution System with Targeted Public Distribution System, (f) establishment of Rural Infrastructural Development Fund and (g) replacement of quantitative controls by tariff.

Similarly, micro credit, routed through self-help groups (SHGs), has emerged as a visible credit channel to the poor as their access to conventional credit channels is constrained by the requirement of collateral and high transaction costs.

While there have been efforts—such as those to promote institutional credit to rural areas, improved access to inputs, subsidies on fertilisers and electricity, minimum support price etc.—these have not been fully successful in protecting the interest of the farmers in general, and small/marginal farmers in particular. Small/marginal farmers have inadequate access to: (a) information, (b) agencies promoting technological change, (c) co-operative networks and related organisations and (d) plethora of governmental interventions. This is significant since, in any case, these farmers are less well placed to take advantage of the markets.

Against the backdrop of the burgeoning demand of population for foodgrains, degrading natural resource base, emerging concerns of climate change and other challenges, the Government has focused on mobilizing higher investment in agriculture, bridging yield gaps that exist across the states/regions, timely and adequate supply of quality inputs, and providing adequate support services to the farmers to make agriculture a remunerative vocation on a sustainable basis. Increasing agricultural production with limited natural resources in a sustainable manner for ensuring food and nutritional security and providing income security to farmers are the major challenges before the Government.

Endnotes

1. The Royal Commission on Irrigation was set up in 1903 and a comprehensive report looked at the options and their prospects for undivided India.
2. The report of the Royal Commission on Agriculture in India submitted in 1928 noted some of the achievements. It noted, "The part which irrigation plays…varies greatly from province to province, but it is of chief importance in Sind, the Punjab, the North-West Frontier Province, Madras, the United Provinces and Bihar and Orissa". Again, in the "Punjab, where the area under irrigation from government works has steadily increased from 2.3 million acres in 1987-88 to an average of 10.4 million acres from 1921-22 to 1925-26, the possibilities of further expansion are far from complete. The Sutlej Valley project will, when completed in 1933-34, provide perennial irrigation for two million acres. But other great schemes are under consideration" (p. 35 of the abridged report).
3. The Royal Commission on Agriculture looked into all aspects of agriculture in their report in 1928. However, a reading of the report shows how nascent agriculture was on that date.
4. Government of India, *Economic Survey,* 2011-12, p. 8.
5. Ibid., p. 9.
6. Government of India, Planning Commission, *An Approach to the Eleventh Five Year Plan*, November 2006, p. 5.
7. Ibid., p. 23.

8. Government of India, Planning Commission, *An Approach to the Twelfth Five Year Plan*, October 2011, p. 7.

3

Phases of Agricultural Development in India

The Economist, in its issue of 16th August 1947, at the time of India's Independence, expressed the hope that 'the energy of the new governments (of both India and Pakistan) will be concentrated without delay on the fundamental question of increasing agricultural production' (cited by Corbridge and Harriss 2000: p. 10). Since Independence, India has made much progress in agriculture. Indian agriculture, which grew at the rate of about 1 percent per annum during the fifty years before Independence, [1] has grown at the rate of about 2.6 percent per annum in the post-Independence era.

Agricultural development in India prior to liberalization in 1991 can be categorized into three phases. The first stage being the reform and consolidation stage of the 1950s and 1960s followed by the green revolution stage of the 1970s and further, the belated green revolution in eastern states along with anti-poverty policy during 1980s. The political economy that succeeded colonial rule in India had the primary task of redistributing resources through a politically guided process of development. During this stage, policies like the abolition of intermediaries, reduction of land revenue and expansion of irrigation facilities were initiated to deal with the institutional problems that were hindering growth and development in this sector (Kalecki, 1972). However, concrete redistributive measures through land reform for agricultural development did not take place as the ruling classes formed alliances with 'dominant propriety classes' consisting of industrial capitalists, the erstwhile *zamindars* and *jotedars* and rich farmers. These groups benefited from the halted and poorly motivated land reform initiatives and also agricultural support and subsidies and the incentives of the green revolution

(Bardhan 1984; Bhaduri 1984). [2]

The simple year to year growth of agricultural GDP at factor cost at constant prices since 1950-51 till 2010-11, can be, at best overall, termed as 'sectoral stasis', to use Byres's terminology. The growth was almost 2-3 percent for the entire period, with usual fluctuations.

However, a significant recovery of growth was observed in the last few years that had pushed the decadal growth rates above 3 percent.

In nutshell, the growth series clearly establish a sharp deceleration in the agricultural sector in the post-reforms period and an unambiguous turnaround during the Eleventh Five Year Plan period.

The trends in aggregate growth rate conceal wide inter-crop differences in growth performance as well as the relative contribution of area and yield changes to output growth.

Five distinct phases of growth can be identified and are outlined below.

3.1 First Phase: Pre-Green Revolution Period (1950-51 to 1967-68)

During the first period after Independence, area expansion which contributed accounted approximately 45 percent to the crop output, mainly consisted of a rise in land use intensity (Sen, Abhijit, 1981). Inter-state inequality in foodgrain productivity remained almost unchanged. This implies that the inter-state disparity in area cropped per person declined, neutralizing the shrinkage in spatial variations in land productivity. The underlying reason as argued by Krishnaji (1993) is a strong negative correlation (-0.82) between land yields and land-labour ratio, which is a macro reflection of inverse size-yield relationship. These observations suggest that the first period agricultural growth had a selectively broad spatial base and, to some extent, succeeded in productively absorbing the rural surplus labour (Rao, Mohan J., 1998).

3.2 Second Phase: Green Revolution Period (1968-69 to 1980-81)

The second phase, in contrast was marked by a sharp decline in contribution of area expansion (less than 18 percent), which was mainly due to lower growth in land-use intensity. However, regional disparities, both in terms of productivity per hectare and per capita output, increased. This implies the absence of compensating changes in the regional pattern of the land-man. They rather worked in opposite direction, because the Green Revolution had a greater spread in states with relatively higher land-man ratios such as north and north-west in comparison to south and east.

3.3 Third Phase: Post-Green Revolution (1981-82 to 1990-91)

The third phase saw an increase in land use intensity and the agricultural growth become broad-based with eastern region picking up significantly. As a result, the macro correlation between land productivity and land-man ratio weakened to -0.7 by the early 1970s and disappeared into insignificance by 1980-83. There was a differential impact of technological change across crops and regions. There is a complex matrix on this participation in the technological change, which works on four dimensions, namely, the crops, region, time period and farmer (Deshpande, R.S., et al., 2004; pp. 185).

This macro finding was ostensibly supported by the results of/from micro level studies of early Green Revolution period showing that there were no diseconomies of scale in land use (Bhalla and Roy 1988, Sampath, 1992); the breakdown of the inverse size-productivity relationship was due to rise in regional disparities, [3] the spatial bade of agricultural growth narrowed considerably, than a reflection of tendencies inherent in the new technology.

It, thus, seems fair to conclude that, while Green Revolution has been successful in increasing the growth through the pure yield component, it has failed to tap the existing potential provided by other sources of growth, namely, increased land use intensities and shifts, both across crops and over regions, in the cropping pattern (Rao, Mohan, J., 1998).

3.4 Fourth Phase: Post-reform Period (1991-92 to 2000-01)

Fourth phase was marked by a period of implementation of the liberalisation policies in agriculture. This had also been the period of a significant slowdown in agricultural growth. Rate of growth of foodgrain production, especially rice and wheat, slowed down significantly. The argument of the proponents of the new policy had been that once terms of trade improve, price incentives would generate a significant supply response. However, the vast literature on the supply responsiveness of farmers has shown that the relationship between prices and output was very weak (Rao, 1998; Sen, 1992; Ghosh, 1992; Nayyar and Sen, 1994; Hazell et al., 1999; Vaidyanathan, 2000). Chand (2004) argued that there was no significant relationship between terms of trade and output growth in Indian agriculture. He observed three sets of patterns of output-price interaction in Indian agriculture in the 1990s: output growth of cereals slowed down while prices were rising; output growth of fruits and vegetables improved when the extent of price rise was small; output growth of oilseeds fell along with a fall in prices.

From these results, Chand argued that "attempts to increase production by raising prices alone do not produce results if efforts on technological fronts, input use, irrigation etc. are ignored". An important outcome of the policy changes in agriculture in the 1990s was the changes in the profitability of cultivation, which varied across crops and regions (Sen, 2004). Using data from the Cost of Cultivation Surveys, Sen (2004) argued that there was a general slowdown in the diffusion of yield-increasing technologies and inputs in the 1990s compared to the 1980s. On the other hand, mechanisation in agriculture grew faster in the 1990s than in the 1980s, though in varying degrees across regions. Sen noted that "the profitability of crop production has been influenced by the pattern of technological change and also by output price movements" (2004, p. 38). To summarise Sen's conclusions:

"During the 1980s, when yield growth was higher and prices of most crops tended to rise faster than the cost of living, the real per hectare margin of GVO over cost...increased for all crops except maize...During the 1990s, with yield growth slowing down for most crops and prices of crops other than cereals and sugarcane rising slower than the cost of living, the real GVO-cost margin fell for most crops other than wheat, sugarcane, barley and tur...Across States also, increases in GVO-cost margins were less evident during the 1990s than during the 1980s (Sen, 2004, p. 38).

According to a Planning Commission review, the sluggish growth in Indian agriculture in the 1990s was mainly due to "weakened support systems", and in particular, "unresponsive agricultural research, nearly broken down extension (and) inadequate seed production, distribution and regulation" (GoI, 2005, p. 197).

3.5 Fifth Phase: Period of Recovery (2001-02 to 2010-11)

The slowdown in growth rate and plateauing of the productivity in major crops was a matter of concern and efforts were needed to step up crop productivity as there was no scope for area expansion. Concerned with slow growth in the agricultural sector during the 1990s, the government focused concentrated attention on agriculture during the decade of 2000s and especially since 2005-06. Public investment in agriculture increased significantly, e.g., gross capital formation in agriculture and allied sectors increased from 13.1 percent of GDP in agriculture in 2004-05 to 20.1 percent in 2010-11 (CSO, 2012).

From pricing support perspective, minimum support prices for many agricultural commodities have been significantly increased. For example, minimum support price of paddy increased from ₹ 570 per quintal in 2005-06 to ₹ 1,080 in 2010-11 and wheat price from ₹ 700 to ₹ 1,285 per quintal during the same period (GoI, 2011). The flow of institutional credit to agriculture and allied sectors increased from ₹ 86,981 crore in 2003-04 to ₹ 4,46,779 crore in 2010-11, at an annual compound growth rate of about 25 percent. The actual achievement in flow of credit exceeded the targets during the period. The government launched two centrally-sponsored scheme: the National Food Security Mission (NFSM) to increase the production of rice, wheat, and pulses by 10, 8, and 2 million tonnes, respectively, by the end of the Eleventh Plan and the Rashtriya Krishi Vikas Yojana (RKVY) in 2007 to incentivize states to increase investment in agriculture and allied sectors, which has led to an increase in allocation for agriculture and allied sectors from 4.88 percent of the State Plan expenditure in 2006-07 to 6.04 percent in 2010-11. In order to develop the horticulture sector, the National Horticulture Mission (NHM) was implemented in 2005-06 (Sharma, 2012).

Due to the concerted efforts, there is a definite growth recovery in the agricultural sector during the last 5-6 years. Foodgrains production growth rate increased to 2.12 percent and the yield growth rate increased to 2.89 percent during the 2000s. The foodgrains production reached a record level of 244.78 million tonnes in 2010-11 and is expected to be over 250 million tonnes in 2011-12, exceeding the target for the year by 5 million tonnes (GoI, 2011).

The pulses production also touched a record of 18 million tonnes in 2010-11. A similar trend was observed in the case of non-foodgrains and all crops. Cotton production in the country increased more than three times from about 100 lakh bales in 2001-02 to 330 lakh bales in 2010-11. In order to exploit the potential of Eastern Plains for enhancing agricultural production, a new programme, "Bringing Green Revolution in Eastern India (BGREI)" [4] has been implemented with an allocation of ₹ 400 crore. However, there is a need to increase allocation under the programme as a large investments required for strengthening the production and market infrastructure in the region.

Endnotes

1. There is scholarly consensus that colonialism was a phenomenon that frustrated all possibilities of growth of the Indian economy. The agrarian sector did not of course remain unchanged under the impact of colonialism but these did not radically change the way production was organized or the techniques that were used (Patnaik, 1999). Over the fifty years that preceded independence in 1947, food grain output in India grew at a meagre 0.1 percent per annum (Blyn, 1966).
2. Resolving the fundamental agrarian question, namely, ending the extreme concentration of land ownership and use and weakening the factors that fostered disincentives in investment and technology adoption, tied workers to a social system with considerable pre-modern features and compressed purchasing power, is dream yet to achieve by the state. The discussion on distributional gains of agricultural growth is the beyond the scope of the thesis. Literature on Indian agriculture is vast and sentiments and opinions expressed range from most enthusiastic to utterly condemnatory.
3. For studies of regional disparities in agricultural growth, see Krishnaji (1975); Bhalla and Tyagi (1989); Bhalla and Singh (2010); Krishnaji (1993); Srivastava (1994); Vaidyanathan (1988); Mashendra Dev (1989).
4. The programme launched in 2010-11 was an initiative taken by the Prime Minster based on the Inter Ministerial Task Force. The BGREI is a part of the Rashtriya Krishi Vikas Yojna (RKVYJ) programme with an outlay of ₹ 400 crore. This scheme was implemented in the eastern region including Assam, Bihar, Chhattisgarh, Jharkhand, Odisha, eastern Uttar Pradesh and West Bengal.

4

Productivity and Efficiency of Indian Agriculture

Understanding Indian agricultural productivity during the last few decades, especially since the early 1990s neo-liberal wave of globalization, is compelling not only for such concerns as food availability and rural poverty but also for the overall development derivation of the Indian economy. The period, 2000-2010, has been, in fact, a turning point in Indian agriculture, as the sector witnessed not only impressive growth rates but also better distribution and democratization of growth among different states of the country.

4.1 Labour-intensive Agriculture

Agricultural production in India is mainly labour-intensive. In the literature, in a labour surplus economy like India, the average labour productivity determines agricultural wage rate (Ranis, 2004). In other words, in the situations where land-man ratios are high and marginal productivity of workers is relatively low, wage determination is based on the principle of sharing output and/or income. This argument is invoked to explain wage formation in the agrarian labour market in South Asia, including India (Bardhan, 1984).

Agriculture, in most developing economies including India, continues to be a core sector, providing livelihood to a significant proportion of the population, especially in rural areas. The sector, in India, has undergone significant structural changes in the form of decrease with a disproportionate retention of high share in the total workforce, but with a fast declining share in the national product. [1] The vast majority of Indian farmers are small and marginal. Their farm size is decreasing further due to population growth. And the quality of the land is deteriorating due to heightened nutrient mining, soil erosion, increasing water scarcity, adverse impacts of climate change and accumulation of toxic elements in soil and water (ICAR, 2010). Land degradation, like climate change, is an anthropogenic induced process and poses biggest threat to sustainable livelihood security of the farming communities across the country.

All of these factors combined with an increased rate of land degradation are contributing towards decline in agricultural productivity leading to food insecurity. Since land resources are finite, requisite measures are required to reclaim degraded and wastelands, so that areas going out of cultivation due to social and economic reasons are replenished by reclaiming these lands and by arresting further loss of production potential. Thus, agriculture in India is imprisoned in the 'vicious cycle' of overwhelmingly low income and disproportionately high unemployment, and plays a significant role in socio-economic development. To overcome this, improving the productive and carrying capacity of agriculture through higher productivity has been an important goal.

4.2 Agriculture Productivity and Rural Incomes

Agricultural productivity growth is closely related to stimulating the rural economies in developing world. [2] Elis (2000) points out that productivity growth in agriculture increases income for the rural communities, which promotes their spending in the non-farm sector. More specifically, agricultural productivity growth will generate more income for the rural population. Further, it is likely to lead the rural farm communities to support non-farm commodities and services such as consumer goods and services, inputs and services to agricultural production and processing and marketing services. In most of the developing economies, agricultural production is the main source of income and a large fraction of the population in the developing world lives in rural areas. Therefore, enhancing agricultural productivity growth has a significant implication for the entire economy of the developing world. Economists, thus, still have a long-standing interest and concern in the study of agricultural productivity. Lynam and Herdt (1989) argued that the appropriate measure of output, which determines sustainability of the crop, cropping system, or farming system is the total factor productivity (TFP). Positive values of the index represent a clear signal of how the output obtained from the system increases more quickly with respect to the input supplied. Non-negative results are therefore an expression of productivity and sustainability of the system (i.e. change in productive capacity of the system). [3]

Banerjee (2008) lists several possible reasons for low use of fertilizer in poor regions: "unwillingness to take risks, the unavailability of credit, lack of the right internal or external incentives for long-term planning, distortions in the land market or a lack of understanding of the benefits of fertilizer use". The diminishing fertiliser use to rice and wheat yield has been observed in case of high yield states. The overall response of the chemical fertilizer use, except for UP, has been negative in the production of wheat in India across the states for the period under study. It is much pronounced in the front-line states like Punjab and Haryana.

Additionally, a generalized additive model (GAM) was tested to derive the marginal response of fertilizer use per hectare to crop yield across the states under the scope of the study. The incremental crop yield to additional unit of fertilizer use per hectare has been substantially lower in high yield states than low yield states. This primarily means that the soil fertility has been lost in these states. The unsustainable substantial increase in the fertiliser use in wheat and rice cultivation has a damaging effect on soil as well as ground water.

4.3 Demand for Workers

Other things being equal, higher output per worker should increase the demand for workers and result in an increase in labour compensation as long as the labour supply curve is not perfectly elastic. Some theoretical models suggest that the causality between the two may also work in the opposite direction. Akerlof (1982), for instance, argued that higher real wages lead to greater effort from workers. Additionally, higher real wages put upward pressure on labour costs and cause firms to substitute capital for labour, thereby increasing the marginal productivity of labour. Further, we also observed the absence of a strong link between the real wage and labour productivity in some states. Bargaining power of workers may be an important factor in weakening the link between labour productivity and real wages. This needs to be investigated further at the micro level as well as for other crops and regions. The deviations of the real wage from the marginal productivity of labour have implications on farms' profitability margins, and thus may also have repercussions for unemployment and employment creation.

India is a vast country with marked regional diversities in agro-climatic environment, resource endowment, and population density. It has been well said by John Augustus Voelcker, an agricultural chemist invited to India in 1892-93, more than a century ago while presenting the report on the improvement of Indian agriculture, and cannot be too often repeated, and ironically still relevant.

4.4 Agricultural Inputs

Indian agriculture still, and continues to suffer from monsoon dependence, fragmented land-holding, low level of input usage, antiquated agronomic practices, lack of technology application and poor rural infrastructure. These are some of the key constraints that stymie a healthy growth. Nonetheless, India has made inspiring strides in the foodgrain production during the last three decades, following the adoption of Green Revolution technology during late 1960s. [4]

Since the 1970s, agricultural productivity growth in the country has been boosted by the arrival of high-yielding varieties, which in combination with fertiliser and irrigation achieved impressive gains in certain areas. The series of interventions [5] initiated in the mid-1960s that led to the Green Revolution in cereals production, transformed the country from one of food deficiency to self-sufficiency. The transformation has taken place since the introduction of new technology in the form of fertiliser responsive high yielding varieties (HYVs) in 1966-67. The increase in the consumption of modern inputs like HYVs seeds, agrochemicals like fertilisers, pesticides, etc., and irrigation as also in the use of agricultural machinery, have been equally steep, breaking the age-old style-hold of traditional and subsistence agriculture.

The progress has been hailed as revolution, popularly known as "Green Revolution". This helped Indian agriculture overcome a "ship-to-mouth" existence and achieve self-sufficiency in production. This achievement was built on a platform of state support; there were price supports, subsidy supports, credit supports and marketing supports. The interventionist role of the state in the 1970s and 1980s led to the creation of a network of institutional support structures in the rural areas. Given the unreformed agrarian economy with dwindling public investment in critical areas like irrigation, the benefits of these support structures were distributed unequally—across crops, classes and regions. Economic reforms after 1991 were based on an explicit rejection of the need to transform the institutional framework of Indian agriculture (Rama Kumar, 2010).

It is also significant that unlike the overall economic growth pattern, agricultural performance in India has been quite volatile. The coefficient of variation (CV) during 2000-01 to 2010-11 was 1.6 compared to 1.1 during 1992-93 to 1999-2000. This is almost six times more than the CV observed in the overall GDP growth of the country, indicating that high and perhaps increasing volatility is a real challenge in agriculture, which is likely to increase in the years to come in the wake of climate change (State of Indian Agriculture, 2011-12).

1.5 Structural Transformation

India is still far away from a turning point in its structural transformation, [6] where the share of agriculture in GDP and in the labour force are starting to converge, and the productivity differential between the non-agricultural and the agricultural sector starts to narrow. Compared to international experience, India's structural transformation has been slow. The share of the agricultural sector in GDP has declined and the remaining non-agriculture sectors, namely, industrial sectors and services have shown growing GDP shares. Absorption of labour in the non-agriculture economy has been slow. Therefore, the difference between the share of agriculture in the economy and its share in the labour force has widened significantly.

At the same time, the accelerating growth of the economy since the 1980s did not lead to an acceleration of the agricultural growth rate. As a consequence of high non-agricultural growth, low agricultural growth, and continued growth of the agricultural labour force, labour productivity in the non-agricultural sector and the agricultural sector has widened at an accelerating rate. Clearly, India is still far away from a turning point in its structural transformation (Binswanger, 2013).

Although estimating TFP is a central technique for understanding trends in productivity, labour productivity also has a useful, independent role to play. In the literature, in a labour surplus economy like India, the average labour productivity determines agricultural wage rate (Ranis, 2004). In other words, in situations where land-man ratios are high and marginal productivity of workers is relatively low, wage determination is based on a principle of sharing output and/or income. This argument is invoked to explain wage formation in the agrarian labour market in South Asia, including India (Bardhan, 1984).

Ever since Solow in the late 1950s (Solow, 1957) put the foundations of the growth accounting methodology, which imply decomposing of total output growth into the contribution of input growth and a residual productivity term, the concept has mushroomed exponentially in popularity. The concept of total factor productivity growth (TFPG) is central in growth accounting, which still remains a dominant concept in growth theory. However, there are still conceptual disputes [7] about the subject and Easterly and Levine (2001) state that "economists need to provide much more shape and substance to the amorphous term TFP". [8]

Productivity growth forms the basis for improvements in real incomes and welfare, and has generated a lot of interest in its own right and is used as a benchmark to rank firms or countries. This has spurred great interest in trying to obtain better and more accurate productivity growth estimates. However, despite being controversial, [9] most economists agree on the importance of productivity growth for an economy. To quote Solomon Fabricant (1961), "Productivity deserves the attention that it has received, for it is a measure of efficiency with which resources are converted into commodities and services that men want. Higher productivity is a means to better levels of economic well-being and greater national strength. Higher productivity is a major source of increment in income over which men bargain and sometimes quarrel. And higher- or lower-productivity affects costs, prices, profits, output, employment, and investment, and thus play a part in business fluctuations, in inflation, and in the rise and decline of industries *and other ingredients of the economy as well* (emphasis added)".

Fundamentally, while productivity is relatively straightforward in concept or simply put, productivity is efficiency in production: how much output is obtained from a given set of inputs. As such, it is typically expressed as an output-input ratio. Thus, productivity growth is, by definition, the difference between the growth of output and that of the factor inputs, each weighted by its factor share. As a simple matter of definition, this is non-controversial. At the same time, a host of measurement issues arise when constructing productivity measures from actual production data.

Endnotes

1. Agriculture was the predominant sector of the Indian economy at the time of Independence, in 1947, with approximately 55 percent share in the gross domestic product and about 70 percent share in the workforce. Contribution of agriculture to the overall GDP has fallen from 30 percent in 1990-91 to 14.5 percent in 2010-11 in recent decades (Status of Indian Agriculture, Government of India, 2012). This decrease, however, in agriculture's contribution to GDP has not been accompanied by a matching reduction in the share of agriculture in employment. About 52 percent of the total workforce is still employed by the farm sector which makes more than half of the Indian population dependent on agriculture for sustenance (NSS 66th Round). In terms of policy response, India announced its National Agriculture Policy (NAP) in July 2000 to prepare agriculture to face the emerging challenges and to help in achieving the goals of growth, sustainability, equity and efficiency. The Policy resolution describes in detail the strategy and policy alternatives which are grouped under the following heads: (i) sustainable agriculture, (ii) food and nutrition security, (ii) generation and transfer of technology, (iv) inputs management, (v) incentive for agriculture, (vi) investment in agriculture, (vii) institutional structure and (viii) risk management.
2. "No country has been able to sustain a rapid transition out of poverty without raising productivity in the agricultural sector" (Timmer, 2007: 5). Empirically, it is proven that those countries most successful at reducing poverty have done so through targeting policies at, and improving, agricultural productivity and growth in the short-term. Long-term poverty reduction is associated more with economic transformation, and this is followed by a decline in agriculture after initial growth (Ahluwalia 1978, 2011). Efforts to boost agricultural productivity historically resulted in more income and wealth and, through the provision of surplus labour and raw materials, the rapid development of industry and manufacturing (Vakil and Brahmananda, 1956; Johnston and Mellor, 1961; Dantwala, 1962; Sen and Ghosh, 1993; Bernstein, 2010; Timmer, 2007;

Lipton, 2006).

3. TFP growth, calculated as ratio of aggregate output to aggregate input, increase without an increase in inputs is the best kind of growth to aim for rather than attaining a certain level of output by increasing inputs, since these inputs are subject to diminishing marginal returns. Thus, in its simplest form: increase in TFP=sustainability, and decrease in TFP= unsustainability (decline in resource base).
4. Green Revolution was associated with several traits that increase the genetic yield potential of crop varieties. For rice and wheat, the yield improvement resulted from a reduction in plant height through the incorporation of genes for short stature. This led to improvements in harvest index (grain-straw ratio) as well as to increases in biomass production. The varieties of wheat and rice available prior to the Green Revolution were tall, leafy with weak stems and had a harvest index of 0.3, i.e. 30 percent grain and 70 percent straw. They could produce a total biomass of 10-12 t/ha. Thus, their maximum yield potential was 4 t/ha. When nitrogenous fertilizer was applied at rates exceeding 40 kg/ha, these varieties tillered profusely, grew excessively tall, lodged early and yielded less than they would with lower fertilizer inputs. The improved varieties on the other hand, have a harvest index of 0.5. Because of their short stature, they are lodging resistant and responsive to fertilizer inputs. Their biomass can be increased to 20 t/ha through fertilizer application. Thus, their maximum yield potential is 10 t/ha. This improvement in harvest index was the single most important architectural change in the rice and wheat varieties that led to doubling of yield potential (Khush, 1995).
5. The interventions were built on three foundations, namely improved package of farming technologies, a system of supply of critical modern inputs and a remunerative price and market environment for farmers. This facilitated the increase in the use of modern inputs like high yielding varieties (HYVs) seeds, agrochemicals like fertilisers, pesticides, etc., assured irrigation, and agricultural machinery to break the age-hold style-hold of traditional and subsistence agriculture. Robert Chambers (1984) eloquently said, "little attention has been paid to the psychology and sociology of ignorance, prejudice and the selective use of evidence in analysing the Green Revolution; yet one obvious feature, with which it is

salutary to start, has been the polarisation between those who have taken views which are positive and optimistic, and those whose views have been negative and pessimistic".

He further cited pessimism expressed by Biplab Dasgupta, who concluded from his study of the Green Revolution in India that some of the major social and economic consequences of the new technology included "proletarianization of the peasantry and a consequent increase in the number and proportion of landless households, growing concentration of land and assets in fewer hands, and widening disparity between the rich and poor households..." (1977, p. 372). Evidence was accumulated of tenants displaced by landlords as agriculture became more profitable, of landless labourers deprived of employment through mechanisation, and of women whose post-harvest employment was destroyed as hand-processing was replaced by mechanical methods. In their negative assessments, some social scientists saw the Green Revolution sharpening social tensions, and some spoke of it turning red.

6. Studies of the patterns of economic growth initiated by Kuznets and later pursued by Chenery, and Johnston and Mellor have shown important regularities in structural composition of economic activity that have recently been reviewed again by Timmer (2009). In their classic paper, The Role of Agriculture in Economic Development, Johnston and Mellor (1961) had argued that as economies develop, the role of the agricultural sector declines in the share of GDP and employment originating in agriculture. Increasing agricultural productivity is crucial in the course of such a structural transformation to assure a steady food supply and to "release" agricultural labour for employment in the non-agricultural sector without increasing food prices and wages confronting the non-agricultural sector. The failure to raise agricultural productivity in the course of modernization would at best slow down, and at worst, halt the process of structural transformation.

   Timmer (2009) has described this phenomenon in terms of the four trends. The rate of decline of agriculture's share in the GDP tends to be faster than the rate of decline of agriculture's share of employment as per capita income increases in the course of modernization. Concurrently, agricultural GDP per worker needs to increase at a rate faster than the rate of growth of agricultural

GDP through technical change for incomes in agriculture to rise as labour is withdrawn from the agricultural sector. Economists in general have been less focused on the backlog of poverty and hunger within the agricultural sector itself, the kind of a phenomenon noted in the areas of slow agricultural growth in India.

7. Lipsey and Carlaw (2001) colligate, "TFP is something close to a scandal that a measurement that is so much relied on for so many purposes seems to be so poorly understood". They delineate into three positions taken by scholars.
   One group holds that changes in TFP measure the rate of technical change [Law (2000); Krugman (1996) and Young (1992)]. We refer to this as the "conventional view".
   The second group holds that TFP measures only the free lunches of technical change, which are mainly associated with externalities and scale effects [Jorgenson (1994) and Griliches (1994)]. We refer to their position as the "J and G view".
   The third group is sceptical that TFP measures anything useful, a measure of our ignorance [Metcalf (1987) and Griliches (1995)].
8. The vast majority of the existing literature on the determinants of TFP faces three main challenges, one defining the term TFP, second, empirically dissecting TFP, and third identifying the policies and institutions most conducive to TFP growth Easterly and Levine (2001, p. 179).
9. Mahadevan (2003), in a controversial paper which led to subsequent argument and counterargument in Felipe and McCombie (2004) and Mahadevan (2004), maintain that the concept, measurement and interpretation of TFP growth remains highly discussed and poorly understood. Two strands of literature, namely, the Cambridge Capital Theory Controversies and the existence of aggregation production function literature, showed that, for all practical (and, indeed, theoretical) purposes, aggregate production functions do not exist. This work has recently been summarized and discussed by Cohen and Harcourt (2003). However, Felipe and Fisher (2003) concluded that the reason for existence of aggregate production function is that, when estimated econometrically, they seem to work (at least at times) in the sense that they yield estimates that are plausible (e.g. the estimated output elasticities are close to the factor shares in national income). Griliches and Mairesse (1997) went

on to say, "production function is estimated as a tool for answering questions which are too interesting to give up even though the framework used may be problematic".

5

Review of Studies on Agricultural Productivity

Indian agriculture has witnessed tremendous changes during the past several decades following the adoption of Green Revolution technology during the late 1960s. The sources and effects of these technical changes on Indian agricultural productivity have been of considerable interest to academicians, policy makers, and politicians. However, the estimates of India's agricultural productivity have been controversial and contentious. Differences in the estimation methods and reliability in the statistics created many debates on the trend of India's agricultural productivity.

Jha and Evenson (1973) estimated the rate of total factor productivity growth (TFPG) in Indian agriculture as a whole to be 0.8 percent per year during the 1956-1961 periods a mere 0.3 percent per year during 1961-1965, but 2.3 percent during 1965-1971. The jump in the TFPG in the latter period clearly reflects the Green Revolution. However, it was too early to conclude definitely.

Easter, et al. (1976), in one of the earlier study at the district level, examined the regional differences in agricultural productivity in two regions, viz., the wheat region comprising selected 73 districts of Punjab, Haryana, Uttar Pradesh (U.P.), Madhya Pradesh (M.P.) and Rajasthan, and the eastern rice region encompassing 69 districts of eastern U.P., Bihar, West Bengal, Odisha, and eastern Andhra Pradesh for the period 1959-60 to 1968-69. They measured the contribution of traditional inputs (land, labour, and fertiliser), irrigation; technology; environmental factors (soil types, rainfall); and infrastructure (transportation, markets).

They estimated the production function by using the

ordinary least squares (OLS) and error component model. They have shown that factors other than the traditional inputs unadjusted for the quality differences are important in explaining agricultural productivity differences within and among regions in India.

In case of wheat region in India, the introduction of new varieties, the expansion of irrigated areas, tube wells and fertiliser led to the substantial increase in production. There is further opportunity to expand irrigation in central and eastern U.P., M.P. and Rajasthan. In the eastern rice region, somewhat different constraints appear to be binding during the period under study. The development and adoption of improved varieties of rice were not widespread. Attention should also be paid to improving roads and related market structures in the region.

Evenson (1985) has used a set of district level data for the two North Indian regions for the period 1959-1975 to empirically examine the question of neutrality of Green Revolution effects on Indian agriculture. He found non-neutral effects of Green Revolution on both output supply and input factor demand. Also, it has favoured the supply of wheat and rice at the expense of other crops on the output side. It has also favoured the demand for fertilizer, tractors, and pesticides, etc. at the expense of bullocks and human labour on the input side.

Productivity gains in Indian agriculture have been quantified and acknowledged in many of the studies. They attributed this growth in TFPG to the application of modern technology and research and extension (R&E) of public and private investment. Rosegrant and Evenson (1992) and Evenson et al. (1999), examined the effects of R&E on productivity in Indian agriculture. Using Törnqvist-Theil index, Evenson et al. (1999) observed that the main sources of agricultural productivity growth in India during 1956-87 were public agricultural R&E; expansion of irrigated area and rural infrastructure and improvements in human capital. Their study also shows that the public benefits from private research can be substantial, indicating that private firms can capture only part of the real value of improved inputs through higher prices.

Private agricultural research accounted for more than 10 percent of growth of TFPG during 1956-87, and in 1966-75, when India was more open to foreign technology, private research contributed 22 percent of productivity growth. Contrary to concerns that TFPG has decreased over time, they found that during 1977-87, the period when the results in regions that adopted HYVs early could be expected to taper off, TFPG was 50 percent higher than before the Green Revolution, indicating that the productivity gains were far from over. The rates of return to public agricultural research were high, and it appears that the government was investing in agricultural research. Expanding public investment in R&E would lead to even greater gains.

The regional breakdown shows that the eastern region had the highest growth in TFP before the Green Revolution but the lowest since. In the North region led by the star performers of Green Revolution states, Punjab and Haryana, the TFP growth over the whole period was the highest, and the southern region performed similarly well during the second or mature phase of Green Revolution, consistent with catching-up after it had been slower to adopt the Green Revolution technology.

In the case of wheat, factor use trends were assessed by Sidhu and Byerlee (1992) for Punjab, between 1972 and 1989, and by Kumar and Mruthyunjaya (1992) for five major Indian wheat producing states (Punjab, Haryana, Uttar Pradesh, Madhya Pradesh and Rajasthan) between 1971 and 1989. These states constitute 80 percent of India's wheat area and just under 90 percent of wheat production; three states (Punjab, Haryana, and Uttar Pradesh) form India's irrigated wheat heartland.

Sidhu and Byerlee (1992) estimated productivity changes for wheat in the Punjab during the 1970s and 1980s. They concluded that despite continuous intensification in input use, productivity gains of 2 percent per year were achieved throughout the period of analysis. These productivity gains can be attributed in roughly equal proportions to the adoption of land-saving technology (modern high yielding varieties (MVs), and fertilizer and to the adoption of labour saving technology (e.g., machinery). The authors warned that future sources of productivity gains capable of increasing TFP at rates equal to those achieved in the past are not evident. Levels of input use were already high, and the gap between the highest and lowest wheat yields obtained by farmers was narrowing (Singh et al., 1987), as is the difference between yields in farmers' fields and yields in the research stations. Any further productivity gains in wheat production, therefore, will depend on more efficient use of inputs, increasing access to information and improving the management skills of farmers.

Kumar and Mruthyunjaya (1992) expanded on the study of Sidhu and Byerlee (1992) by considering wheat producing regions in India other than the Punjab. During the period 1970-89, they found that TFP indices grew annually by 1.9 percent in Punjab, 2.7 percent in Haryana and Rajasthan, 2.6 percent in Uttar Pradesh, and 0.4 percent in Madhya Pradesh. The authors also decomposed the TFP estimates to see the sources of growth and found that market infrastructure, mechanization, and research were the most important ones. They found that in all the states studied, the "modern" inputs of fertilizer and machine power grew during 1979-1989 at even faster rates in the irrigated heartland.

Pesticide use rose sharply from a limited base in Punjab and Haryana, while applications were negligible in the other states. Human labour fell in all states, particularly in Rajasthan and Madhya Pradesh, and animal labour fell precipitously in all states. Seed rates (kg/ha) rose in Punjab and Uttar Pradesh, rose slightly in Haryana, and showed no trend in the other states. Manure use in wheat production fell in all states except Madhya Pradesh. They concluded that further productivity gains in wheat can be achieved through more efficient use of inputs such as fertilizer, water, and machines. This conclusion is consistent with the results of Sidhu and Byerlee (1992). Since research was an important source of growth, the authors also indicated that new varieties which can break the current yield ceilings will be instrumental in the future productivity of Indian wheat farmers.

A policy brief by Kumar et al. (1995) concluded that TFP growth for wheat in India dropped from 1.4 percent per year during the 1970s to 1.1 percent per year during 1980s. This decline was attributed to the declining real investments in agriculture, such as research. Projected wheat supply in the future was estimated to meet demand in the long-term so long as productivity through public investment in irrigation, infrastructure development, research and efficient use of water and plant nutrients are maintained. These recommendations were also consistent with the studies above.

Dholakia and Dholakia (1993) estimated the sources of growth of Indian agriculture for the sub-periods:. 1950-51 to 1966-67 (pre-Green Revolution period), 1966-67 to 1980-81 (initial phase of Green Revolution) and 1980-81 to 1988-89 (modernisation phase). They found that the contribution of TFPG to agricultural output growth declined during 1980-89. The other notable study in this regard is by Fan, Hazell and Thorat (1998) which estimated TFP for agriculture at state-level using Törnqvist-Theil index for the period 1970-1994. The annual compound growth rate of TFP indicated that TFP as estimated by Dholakia and Dholakia (1993) increased at the rate of 1.77 percent per annum during 1967-68 to 1977-78 as against 1.73 percent per annum during 1978-79 to 1988-89. In contrast, Tripathi and Prasad (2008) found that Indian agricultural growth relied almost entirely on increase in conventional factors while growth in productivity was negative. They used time series data related to broad agricultural outputs which included farming, livestock, forestry, and fisheries and 3 conventional inputs: labour, land, and capital, to construct an index of TFP for the period 1969-70 to 2005-06. The TFP so constructed, suffered from theoretical limitations and aggregations.

However, in the pre-Green Revolution period of Indian agriculture during 1952-53 to 1964-65, TFP had grown only at the rate of 0.53 percent per annum. As per the estimates of Fan, Hazell and Thorat (1999), the growth in TFP worked out to be at the rate of 1.39 percent per annum during 1970-71 to 1980-81, as against 1.36 percent per annum during 1981-82 to 1990-91. However, their study showed that in the early years of economic reforms (for the years 1991-92 to 1994-95), the TFP growth of Indian agriculture was 2.67 percent per annum. The study also reported state-level estimates for the period 1970 to 1994. The states with a TFP growth rate in the range 0-1 percent per annum were Andhra Pradesh, Karnataka, Uttar Pradesh, Himachal Pradesh and Kerala; with TFP growth rate greater than 1 were Punjab, Bihar, Odisha, Maharashtra, West Bengal and J&K. The states with negative TFP growth were Haryana, Madhya Pradesh, Gujarat, Assam and Rajasthan. In a recent study by Nin-Pratt et al. (2010), while comparing TFP growth rate of China and India, they found almost zero TFP growth throughout 1961 to 2006. They did not find any structural change in India's agricultural TFP series. The result was in total contradiction with the majority of the findings on Indian agriculture.

The study found that TFPG has contributed significantly to the acceleration of overall agriculture, facilitating release of scarce resources from agriculture to the other sectors in the economy. Thus, TFPG has been the driving force behind the acceleration of overall growth in the Indian economy achieved during the 1980s. The main determinant of TFPG has been found to be the use of modern inputs like fertilizers, HYV seeds and irrigation.

The empirical study by Sivasubramonian (2004) has shown the trend of growth rate of TFP in Indian agriculture, which was recorded at 1.65 percent during 1950-50 to 1960-61, which declined to 0.88 percent during 1960-61 to 1970-71 and further declined to -0.35 percent during 1970-71 to 1980-

81. Further, during 1980-81 to 1990-91, the growth rate of TFP increased to 1.89 percent and then again declined to 1.68 percent during the period 1990-91 to 1999-2000. However, all the three studies show that the growth rate of TFP in Indian agriculture has fallen during the 1970s. Considering the share of TFP in agricultural GDP growth, as estimated by Sivasubramonian (2004), TFP growth has contributed 54.5 percent of agricultural GDP growth during 1950-50 to 1960-61 and then the share declined to 38.1 percent during 1960-61 to 1970-71 and -23.3 percent during 1970-71 to 1980-81. However, the share increased to 55.1 percent during 1980-81 to 1990-91 and then to 56.6 percent during 1990-91 to 1999-2000. Thus, it is obvious that, during the last couple of decades the agricultural GDP growth is largely explained by the TFP growth, although the growth rate of TFP has declined during the 1990s as compared to the 1980s.

Kumar and Rosegrant (1994) assessed the TFPG in different regions of India, and estimated the sources of productivity growth in rice for the period 1970-71 to 1988-89. They found that the TFP index has risen by around 1.85 percent annually in the southern region, 0.76 percent in the northern and 0.36 percent in the eastern region. In the western region, due to wide fluctuations in weather, wide variation in the TFP index was observed and the estimated annual growth was negative (-0.98), but insignificant. They observed that gains in rice production essentially came from the improved utilization of available infrastructure and from the resulting increase in yield per unit. They also found very high marginal returns to public investment in rice research.

Desai (1994) used a comprehensive set of inputs that incorporated 11 farm inputs which included, both credit financed-land, labour, seeds, organic manure, fertilizers, pesticides, diesel electricity, irrigation charges and public and private capital—consisting of land improvements, farm equipment and tools, public and private machinery, livestock and inventories. For the period 1950-51 to 1989-90, he estimated that a close to 38 percent of the growth in output of agriculture and allied sectors was accounted for by technical change.

This estimate was lower than that of Rosegrant and Evenson (1992) who had little over 50 percent. Both studies, however, show that the contribution of technical change was in land, labour and intermediate inputs.

Desai and Namboodiri (1997) considered both the price and non-price factors to understand the process of change in TFPG in Indian agriculture for the period 1966-67 to 1989-90. They observed that the technical change is a superior strategy compared to either extensive farming or intensive farming as it, unlike other two options, increases production at reduced unit-cost prices in real terms which benefits poor the most.

The technical change, they concluded, is mostly influenced by the government expenditure on R&E, followed by liberal tenancy regulations, rural literacy, and land distribution through existing ceilings, rural roads, marketing and banking infrastructure, balanced use of fertilizers combined with efficiently managed canal irrigation in the order of importance. Their findings supported the earlier findings.

Mukherjee and Kuroda (2001) used the Törnqvist-Theil methodology to construct the TFP index for Indian agriculture in fourteen states from 1973 to 1993. They found the TFP index to be 1.73 for 1973-79, 2.51 for 1980-89, 1.34 for 1990-1993 and 2.19 for entire period 1973-2003. Bosworth and Collins (2007) used growth accounting approach and estimated TFP growth in primary sector for India to be around 0.8 percent during 1978-2004, 1 percent for the period 1978-1993 and 0.5 percent for the period 1993-2004.

An analysis of productivity of the crop sector in the Indo-Gangetic Plains (IGP) by Kumar et al. (2004), revealed that the TFPG of the crop sector in the IGP had risen at the rate of 1.2 percent per annum during the period 1980-81 to 1996-97. The TFP results for different agro-eco-regions also showed considerable variations. The Low-Gangetic Plain (LGP) region depicted the highest growth in TFP (3.1 percent) and MGP, the lowest (0.37 percent). The TFP growth rates were estimated at 1.4 percent in the Trans-Gangetic Plain (TGP) and 0.9 percent in the Upper-Gangetic Plains (UGP). In IGP, one-third of output growth was contributed by the TFP. However, the contribution of TFP to output growth varied from as high as 57 percent in the LGP to a meagre 17.3 percent in the MGP.

The share of TFP in the output growth of the crop sector in the TGP and the UGP regions was observed to be 34 percent and 26 percent respectively. The output growth in the UGP and the MGP was input-based, while in the LGP, it was technology-based. The output growth in the TGP was input- as well as technology-based. The analysis confirmed that contribution of TFPG to output growth had started declining and was, in fact, showing a tendency of further deterioration in the process. Productivity growth, which picked up during the early-1980s, could not sustain during 1990s and this situation raised an alarm for the policymakers and researchers of the country.

Kumar and Mittal (2006) estimated crop-specific TFP growth across different states for paddy and wheat. They found that the TFP of paddy started showing deceleration in Haryana and Punjab but the TFP of wheat was still growing in these two Green Revolution states. About 60 percent of the area under coarse cereals was facing stagnated TFP. Similarly, the productivity gains which occurred for pulses and sugarcane during the early years of Green Revolution had exhausted their potential.

However, the above mentioned studies of TFPG in Indian agriculture have not distinguished the increment in productivity that occurred from technical progress and from that which resulted from improved technical efficiency in the application of already established technologies. This is because the technically efficient production can be achieved if farmers follow the best practice to apply the technology. To the extent that farmers do not produce with technical efficiency due to differences in their capacity to use new technological knowledge, technical progress is not the only source of total factor productivity growth. Changes in productivity arise from two connected parts: technical progress and changes in efficiency. Hence, the decomposition of total factor productivity growth into technical progress and changes in efficiency provides more information about the application of production technology. From a policy point of view, this decomposition is important because without using the existing technology to its full potential it may not be meaningful to embark on the introduction of new technologies.

The earliest stochastic frontier function study using Indian data appears to be one by Kalirajan (1981). The author, using a random sample of farms, explored technical efficiency in paddy production in Tamil Nadu by estimating a Cobb-Douglas production function through maximum likelihood. The average technical efficiency was 67 percent. The author also showed that management practices and contacts with local extension agents had a significant positive impact on technical efficiency.

Huang and Bagi (1984) examined the technical efficiency of a sample of 151 farms in two of the Green Revolution star states viz., Punjab and Haryana. Their analysis, based on a cross-sectional translog production frontier via maximum likelihood, showed an average technical efficiency level close to 90 percent, while the performance of small vis-à-vis large farms was almost equal.

Kalirajan and Shand (1985) estimated a Cobb-Douglas production frontier by maximum likelihood for a random sample of 91 paddy farmers from the Coimbatore district of Tamil Nadu. In a second step analysis, where farm level technical efficiency was dependent variable, they found that the level of schooling was not statistically significant in explaining differences between maximum and actual yields. However, the farmers' non-formal education, defined as their understanding of current technology, had a significant positive impact on productivity.

Ray (1985) used linear programming to measure efficiency for a sample of 63 West Bengal farms. He decomposed efficiency measures into output or technical efficiency and informational efficiency. The latter, he defined as the ratio between optimal output given the existing technology and optimal output when additional information is available. Univariate and multivariate statistical tests were conducted to compare the performance of three farm groups classified according to size. The results revealed that informational efficiency was very low for small farms. However, there was no significant difference in output efficiency across farm size groups. He concluded that marked improvements could be attained by the diffusion of information about the standard crop production technology.

Kalirajan and Shand (1997) decomposed output growth into input growth, [1] technical change and efficiency change for the period 1980-83, 1984-87, and 1988-90, and estimated technical change as a shift in production frontiers.

Allowing technical progress to be non-neutral, they estimated TFPG, the stochastic varying coefficient frontier using the Cobb-Douglas technology function for the period 1980-90. Their analysis shows that TFPG was negative in 4 out of the 15 states during 1980-83 and that, by the end of decade, it was small for those states where the contribution of TFPG was positive. While the input growth contributed more than 50 percent to the output growth, the contribution of technology to output growth declined substantially, particularly from 1988 to 1990. The share of fertilizer and electricity in the consumption of core inputs, which enjoy heavy subsidies in Indian agriculture, increased from 16.8 percent in the 1970s to 29.2 percent in the 1980s (Misra and Hazell, 1996).

An average of only around 18 percent could be attributed to technological changes in the pre-reforms period but more to gains in technical efficiency. Importantly, the contribution of increasing technical efficiency to output growth remained more or less at the same levels in most states in the pre-reforms periods. Thus, since the introduction of the HYVP, Indian agriculture experienced low rates of technological progress together with negligible improvements in technical efficiency, and output growth in the sector became increasingly dependent on input growth. Here, TFP can be obtained by adding the technical change and technical efficiency.

Second, the deterioration of land infrastructure, particularly the existing water conservation systems, exerted a constraint on research that is generally irrigation-oriented. Gross fixed capital formation in agriculture at 1980 constant prices sharply declined to an average annual growth rate of 1 percent during 1980-90 from a corresponding growth rate of 5 percent during 1970-80. Also, constitutionally, agriculture is under the control of States and with increasing industrial lobbying for protection and farmers lobbying for subsidies at State levels poses additional problems for agricultural growth in India. Thus, the inference is that the agricultural institutions across States would not have significantly contributed to widening regional income inequality over and above the existing inequality due to initial differences in factor endowments across the States. Their results, obviously, suffer the problem of excessive aggregation. It would be desirable to conduct estimation at a lower level of aggregation, given India's big size and enormous diversity.

A significant development in the econometric approach was achieved by combining the developments in duality theory and flexible functional forms with econometric theory to economic efficiency. The methodology was introduced in Berndt and Christensen's (1973) seminal paper on the translog production function. This methodology involves specifying a function representing the technology (a production, cost, or profit function) and econometrically estimating it or its derivatives. [2] Paul and Mehta (1991), for the first time, applied translog cost function in Indian agriculture for the period 1960-61 to 1982-83 to test the hypotheses regarding the structure of production technology. They estimated the factor demand, elasticity of substitution, and biases of technical change. They used two outputs: crop output and livestock production and four inputs: labour, land, fertilizer, and other inputs. Their cost share equations involved the logarithm of prices of four inputs and the logarithm of output and time. By dropping one of the share equations, they estimated the simultaneous equation system by the iterative non-linear Zellner method yielding maximum likelihood estimates.

The data set used in the study supported the null hypothesis of non-homothetic cost function showing a well behaved production structure. The coefficients of time in labour and capital cost share equations were positive and significant. This implies that technical change had been biased towards the use of labour and capital. These biases, they argued, cannot be explained in terms of a price-induced hypothesis: 'they only indicate the presence of biased innovation possibilities in Indian agriculture' (p. 536). They also reported a neutral technical change to land use, fertilizers and other inputs. However, there are problems with this study. The problem is two-fold: first, as there are wide variations in the nature of technical change among different crops, so treating entire crop output and livestock production as the output measure is questionable. Second, relates to the input category of fertilizer and other inputs, which are heterogeneous. Getting meaningful measures of factor prices and viewing them as exogenous variables, especially at the macro level is tricky. This approach might be meaningful at the crop level, but at the macro level, the prices are generally endogenous.

Jha and Rhodes (1999) estimated stochastic frontier production function for wheat in two Indian States for the period 1981-82 to 1982-83: Haryana (which has been significantly affected by the Green Revolution) and Madhya Pradesh (which has been lagging behind). Their paper attempted to ascertain the requirements (in terms of ownership of factors of production) for successful adaptation to the Green

Revolution. They found that in Haryana, but not in Madhya Pradesh, larger farm size and ownership of land and machines positively influenced technical efficiency and, concluded that, with the Green Revolution advancing, land consolidation and vesting of clear ownership rights of land and capital with farmers becomes important. However, the period under study was too small to conclude significantly.

There is another approach (Färe, et al., 1994a), which estimates the non-parametric Malmquist Productivity Index (MPI) and provides an interesting application of a modern approach to the measurement of productivity change. The approach uses linear programming techniques to construct a piece-wise linear production frontier. The observed output-input combinations are then compared to this production frontier through distance functions (Cones, 1992) which are, subsequently, used for the calculation of a Malmquist-type index of productivity changes. One important advantage of this method is that it uses only data on input and output quantities to construct the frontier.

In a cross-country study, Arnade (1998) used the data envelopment analysis (DEA) method to calculate the Malmquist Productivity Indices (MPI) for the agricultural sectors of 70 countries, which includes India, from 1961 to 1993. He found that in the developed countries, agriculture generally remained efficient and productivity had improved. However, in many developing countries, including India and Brazil, agriculture was relatively inefficient and multi-factor productivity had fallen. The agricultural growth in these countries appeared to originate from the growth in substantial input use of machinery and fertilizer rather than from productivity gains. In case of India, the output grew by 2.55 percent and inputs like tractors and fertilizer use grew by 11.55 percent and 8.67 percent respectively. But the productivity showed a negative growth of -2.60 percent during 1961-1993.

Instead of using output distance functions, Arnade constructed MPI using input distance functions which minimize the input requirements for given outputs, which seems an unreasonable assumption for developing countries' agriculture. This is due to the fact that farmers in developing countries try to maximize the output with the given fixed quantities of inputs. In this case, output orientation measure could be more appropriate.

Murgai, et al. (2001) attempted to determine the long-term productivity and sustainability of irrigated agriculture in the Indian and Pakistan Punjab by measuring trends in total factor productivity for production systems in both the states since the advent of the Green Revolution. The period, they considered for Indian and Pakistan Punjab were 1961-94 and 1966-94 respectively. They, however, reported that estimates of TFP growth by using conventional techniques (Divisia and Törnqvist-Theil indexing) were underestimated by 100-300 percent because of serious 'bias' in the conventional TFP measures. Using Törnquist-Theil index to measure total factor productivity, their major findings were: first, despite a higher output growth (5.5 percent and 3.2 percent in Indian and Pakistan Punjab respectively) and crop yields in Indian Punjab, productivity growth (1.9 percent and 1.3 percent in Indian and Pakistan Punjab respectively) was higher by a small margin.

Moreover, the lowest growth in productivity took place during the initial Green Revolution period (as opposed to the later intensification and post-Green Revolution period) and in the rice-wheat system in both the states. The time lag between adoption of Green Revolution technologies and realisation of productivity gains was related to learning-induced efficiency gains, better utilisation capital investments over time, and to the downward measurement bias and authors attribute, in the methodology. Second, input growth (3.0 percent and 1.9 percent in Indian and Pakistan Punjab respectively) accounted for most of the output growth in both of the Punjabs during the period under study. Third, intensification, especially in the wheat-rice system, resulted in resource degradation in both Punjabs. Data from Pakistan Punjab showed that resource degradation, particularly soil and water degradation reduced overall productivity growth from technical change and from education and infrastructure investment by one-third.

There seems to be wide significant variations in India's agricultural TFP growth estimates, even among similar methods and datasets, particularly using FAO data.

Coelli and Rao (2005) and Nin-Pratt, et al. (2010) estimated TFP using Malmquist indexes. Their results differed: Nin-Pratt, et al.'s estimates indicating growth to be half of what Coelli and Rao estimated. Because both Malmquist estimates were extracted from broader, global analysis, individual country estimates may have been affected by the dimensionality issue or the number of commodities and countries included in the analysis (Lusigi and Thirtle, 1997). Moreover, Coelli and Rao (2005) noted that if shadow prices are indeed correctly estimated, for many countries the estimates may significantly differ from the sample average due to country-specific factor abundance or scarcity. Indeed, these two issues could be the reasons why the TFP growth estimates from Coelli and Rao differed by method with their Törnqvist index (using Malmquist-derived cost shares) indicating a slower TFP growth than does their Malmquist index.

It is also interesting to note that the three Törnqvist approaches using FAO data 8 not only examined the same time periods but also generated widely different TFP growth estimates. Coelli and Rao's (2005) average annual Törnqvist estimate of 0.90 percent was less than Fuglie's (2012) estimate of 1.39 percent per annum TFP growth, which itself was significantly lower than Avila and Evenson's (2010) estimate of 2.41 percent per annum. Differences between them must have lied in factor quality adjustments and the cost shares employed. Fuglie's quality adjustments for land inputs were to account for irrigation investments thereby lowering the TFP growth in countries with substantial investments in irrigated cropland, such as India. The three studies' respective cost shares weighting factor growth also differed: Coelli and Rao employed shadow cost shares from their Malmquist index, Avila and Evenson sourced their cost shares from Evenson and Kislev (1975) and Fuglie's sourced it from Evenson, Pray, and Rosegrant (1999).

Bhushan (2005, 2006) estimated the total factor productivity growth of wheat in India for the post-Green Revolution period. He used data envelopment analysis to estimate Malmquist TFP index for major wheat producing states in India viz., Punjab, Haryana, Madhya Pradesh, Uttar Pradesh and Rajasthan. He found TFP growth rate to be the highest in Punjab and Haryana which was attributed to technical progress in these two states. Rajasthan (with no efficiency change) and Uttar Pradesh (with improvement in efficiency and negative growth in technological progress) had positive TFP growth rate while Madhya Pradesh (no change in efficiency and negative growth of technical progress) reported negative TFP growth rate. As compared to 1980s, mean growth of TFP was found to be higher in 1990s and the primary source of TFP growth was technical progress and not efficiency improvements. In any case, further study is needed in order to understand more about the remarkable technical progress in Indian wheat production.

However, the major challenge to Indian wheat production is the decline in technical efficiency. Poor performance in technical efficiency in major wheat producing states like U.P. indicates a great potential to increase productivity through improved technical efficiency. The empirical result of this study suggests that the efficiency level had not been maintained. The decline in efficiency in fact had eroded part of the positive impact of the technical progress. For wheat production growth to sustain in the future, the Indian government might need to look more carefully into the factors that have caused such a serious decline in efficiency.

The discussion above suggests that cereal yields have continued to rise on an average across Asia since the Green Revolution era, but annual growth rates are slowing (Rosegrant and Hazell, 2000; Hazell 2008). This is confirmed by more careful, micro-based studies of wheat and rice yields in the Indo-Gangetic Plain (Murgai et al., 2001; Ladha et al., 2003; Cassman and Pingali, 1993; Bhandari et al., 2003) of India's major irrigated rice growing states (Janaiah et al., 2005) and in East Asia's rice bowls (Pingali et al., 1997).

Cassman and Pingali (1994) also calculated declining productivity indices in the rice-wheat systems of the Indian Punjab. According to their estimates, TFP steadily increased during the 1970s and levelled for a while before slowly dropping. Both studies attributed the decline in TFP to deterioration of agricultural resources, especially the soil. Clearly, proper management of agricultural resources is also essential in the medium- to long-term productivity of wheat based cropping systems.

Both Bhalla and Singh (2001) and Kumar (2001) in their studies observed that a major part (almost 50 percent) growth in the crop production in Indian agriculture was due to the increased use of modern inputs and R&E. But the realization of gains from use of modern inputs, however, was constrained by diseases and pests, and by inefficiencies in farm production. The waterlogging and salinisation were the most serious problems in this regard (Bhatia, 1988; Dhawan, 1988). This may be due to the unsound agricultural production practices, which have been encouraged by irrigation water management and pricing policies of the government.

Endnotes

1. The output consists of crop production, livestock, and forestry, whereas inputs, they taken are: land (sown acreage and pasture areas), labour (workers in cropping sector only), machinery (includes tractors and draft animals and measured in horse power), and chemical fertilizers (gross weight of nitrogen, phosphorous, and potash).
2. Antle and Capalbo (1988) provide the theoretical underpinning.

6

Environmental Sustainability of Indian Agriculture

6.1 Environment and Development: The Conflict

Natural and environmental resources are the basis of all economic activities. Soil, forests, mines, water, air and other natural resources are productive assets of an economy. In fact, productivity of an economic system depends, to a large extent, on the supply and quality of its natural resources. Poor quality of water and air adversely affects the health of population, causing reduced labour productivity and premature mortality. Similarly, soil degradation and erosion leads to reduced and inferior quality of agricultural output.

Economic development often depends on the utilisation and conversion of natural resources into useful goods and services. This process of consumption and conversion uses energy, minerals and other natural resources such as air, water, land and bio-diversity. It also produces wastes which are released into environment with adverse impacts.

Economic activities, in turn, affect the quantity and quality of natural and environmental resources. Mining, lumbering, manufacturing, fishing and a host of other economic activities change the stock of natural resources which calls for appropriate trade-off between the needs of present and future generations.

Increase in economic activities—due to growing population and/or enhanced consumption levels—often impacts environment negatively. Degradation of natural resources and losses in environmental quality adversely affect people at large. As a consequence of high level of economic activities, emissions, effluents and waste discharges exceed the carrying capacity of different resources. Economic growth without environmental considerations can cause serious damage to the quality of life of the present and future generations. It is in this context that economists distinguish between economic growth and economic development. Economic growth means sustained increase in per capita real income. Economic development is a much wider term. It includes economic growth and at the same time encompasses questions regarding patterns of production, distribution of national income, consumption behaviour of the people and concern for environment.

Economic progress improves our standard of living and makes our life more comfortable. On the other hand, it is this very progress that can lead to degradation of the environment. Any increase in national income would arise only from increased production of goods and services involving greater consumption of resources such as land, forest, fuels etc. whose supply is, essentially, limited. While some of these resources may be renewable, others get depleted and, ultimately, exhausted with continuous use. Any attempt to preserve the non-renewable resources might require compromising on the growth rates.

The conflict between economic development and environment is, essentially, a conflict between short- and long-run priorities, between the interests of current and future generations. Large exploitation of natural resources today, while increasing the economic growth for the current generations, would lead to gradual exhaustion and degradation of these resources, thereby reducing their availability to our future generations and adversely affecting their output, income and living standards.

6.2 Environmental Pollution

Demographic expansion, over exploitation of natural resources, mega development projects—with least concern for sustainability and conservation—have greatly affected the environment. The magnitude and alterations that humanity can, make to environment, threaten to undermine the basic natural process that sustains biosphere. The process of economic development necessarily entails exploitation of natural resources. However, when such exploitation becomes indiscriminate and is undertaken without adequate environmental safeguards, the consequences can be disastrous.

Pollution can be defined as an undesirable change in the physical, chemical, or biological characteristics of the air, water or land that can harmfully affect health and survival activities of human beings or other living organisms. Pollution refers essentially to a process by which a resource is rendered unfit for some beneficial use. Of the various kinds of pollution (air, water, land, noise, radiation and odour) that affect the quality of life in India, water pollution is by far the most serious in its implications for the health and well-being of the people.

Environmental pollution is one of the major problems faced by the world community, especially in the cities of the developing countries which have experienced unbridled growth of population, urbanization and industrialization. Municipal services—such as water supply and sanitation, drainage of storm water, management of solid and hazardous wastes, supply of adequate and safe food and housing—are unable to keep pace with urban growth. The unplanned location of industries in and around urban areas and volume of traffic have caused serious pollution problems. All these factors have led to deteriorating environmental conditions, adversely impacting the health of the people.

Pollution has become a major threat to the very existence of mankind on this earth. Pollution of various resources has gone to such an extent that people are unable to breathe fresh air and drink fresh water. True, advancements in science and technology have added to human comforts by way of automobiles, electrical appliances, supersonic jets, modern medicines, communications and entertainment. However, these blessings have been diluted by serious problems posed by ecological imbalances and climate changes.

Environmental degradation not only affects us but also has repercussions for future generations. Globally, the need for sustainable development is being emphasized. Sustainable development attempts to strike a balance between the demands of economic development and the need for protection of our natural environment.

6.3 Sustainable Development: The Balancing Act

The concept of sustainable development is inextricably linked with environment protection. It is a strategy for improving the quality of human life while living within the carrying capacity of supporting ecosystem. It is clear from the above definition that sustainable development entails commitment, protection and preservation.

Sustainable development is that which meets the needs of the present without compromising the ability of future generations. The heaviest burden in international economic adjustment has been carried by the world's poorest population. The consequence has been the considerable increase in human distress and over exploitation of land and water resources to ensure survival in the short-term. Further, the desire to solve all social and human problems at one stroke has influenced the pace of development. Its reaction is visible in the atmosphere and in natural, physical and social environment. Awareness regarding the consequences of this event has developed into a consensus referred to as ecological concern.

Sustainable development advocates economic progress in an environmentally responsible manner. Sustainable development attempts to strike a balance between the requirements of economic development and the need for protection of the environment. It seeks to combine the elements of economic efficiency, inter-generational equity, social concerns and environmental protection. Although, the term sustainable development has many interpretations, it generally refers to non-declining human well-being over time.

Economic growth and development have to be guided by the compulsion of sustainability, because no country can afford to ignore the economic as well as the environmental threat that a fast-deteriorating ecosystem poses to its economy. No country is immune to the reality of climate change, ecological degradation, depletion of the ozone layer and contamination of fresh water.

Sustainable development is a difficult balancing act in countries with low incomes. Society has to simultaneously accomplish three things with trade-offs: improve economic well-being with social justice for the present generation, yet manage with more restrained use of land, air, forest, energy, and water resources, and protect future generations. The choices are more difficult in developing countries because they affect people's livelihoods. Such a 'stewardship' to succeed therefore needs to respond to people's needs, share information on choices and costs, and ensure participation and ownership.

6.4 Agriculture-Environment Interface

Agriculture both causes and suffers from environmental degradation. The International Soil Reference and Information Centre (ISRIC) in Netherlands, estimates that every year, 3 million hectares (mha.) of cropland is mired by soil erosion, 4 mha. is turned into deserts, and 8 mha. is converted into non-agricultural uses such as homes, highways, shopping centres, factories, reservoirs, etc. Over the past 50 years, some 1.9 billion hectares of agricultural land (an area greater than that now in production) have been degraded to some extent. The causes and consequences of degradation as observed by Tilman et al. (2001) and Ruttan (2002) vary, which include soil loss and degradation; water-logging and salinity; the co-evolution of pests, pathogens, and hosts; and the impacts of climate change.

6.5 Climate Change and Agriculture

Climate change is a serious global environmental concern, which is primarily caused by the building up of greenhouse gases in the atmosphere. Recent discourse on climate change has underscored the fact that climate change occupies a high priority on the environmental agenda of international community.

Global climate change due to rising levels of green house gases (GHGs) in the atmosphere is one of the most serious environmental concerns of present time. The Intergovernmental Panel on Climate Change (IPCC), established in 1988 by the World Meteorological Organization (WMO) and the United Nations Environment Programme (UNEP), has worked extensively on evaluating past trends and the future prospects of climate change. The Fourth Assessment Report (AR4) of the Panel was released in November 2007. The Panel has concluded that the fact of global warming is unequivocal and there is enough evidence to indicate that this is due to anthropogenic reasons. Recent years have been the warmest since 1860, the year from when regular instrumental records are available. It is important to note that global warming will affect mankind seriously.

Since global warming is caused by total accumulation of GHGs and each country's emissions add to the stock, the cost of each country's emissions is borne by the entire global community. Similarly, the benefits of reduction in emission by any one country accrue not to the country itself but to the entire global community. This is classically a situation characterized by externalities in which individual countries will not have a sufficient incentive to undertake mitigation unless all countries co-operate in what is felt to be a collective effort with an equitable distribution of the burden of reducing emissions.

The accumulated scientific evidence clearly indicates that if the rate of GHG emissions is not reduced very significantly, there will be a damaging effect on the global climate.

According to IPCC, the global temperature may rise by 2-4.5 degree centigrade by the end of this century, including a 2.7-4.3°C increase over India by 2080s if adequate actions are not taken to reduce emissions globally. At the national level, the observed increase in surface air temperatures over the past century is 0.4 degrees centigrade.

Changes observed in the IPCC Assessment Reports

include: (a) shrinkage of glaciers, (b) thawing of permafrost, (c) lengthening of mid-to-high latitude growing seasons, (d) poleward and altitudinal shifts of plant and animal ranges, (e) decline in population of some plants and animals, (f) earlier flowering of trees and (g) changes in emergence of insects, and egg-laying in birds. Associations between changes in regional temperatures and the observed changes in physical and biological systems have been documented in many aquatic, terrestrial and marine environments.

Changing environmental attributes are sure to affect the species spectrum and depending upon the pace of change, the profile and composition of forests is also likely to change. In arid or semi-arid areas, productivity is expected to decrease. Trends of drying up of wetlands and degradation of ecosystems have been observed.

The water cycle is totally dependent on the temperature regime. Increasing global temperature and resultant faster retreat of most glaciers is expected to affect the snow-fed perennial water regimes. Accordingly, the vegetation status of catchments, magnitude and frequency of floods, precipitation, run-off and groundwater recharge may be affected. It may substantially affect hydropower generation and increase in irrigation withdrawal, due to higher crop evaporative demand.

Estimates of crop yield responses to climate change vary widely, depending upon species, cultivation practices, soil properties, and direct effects of carbon dioxide (CO_2) on plants. Even though increased $C0_2$ concentration can stimulate crop growth and yield, the benefits may not always overcome the adverse effects of excessive heat and drought.

In countries with predominantly agrarian economies, increase in temperature and reduction in precipitation could—in the absence of adequate irrigation and related infrastructural interventions—dampen the economic growth by reducing agricultural productivity. These impacts can be moderated by efforts at adaptation but these are expected to be costly and will impose a burden on both producers and consumers.

For South Asia, warming has been predicted to be above the global mean and in Southeast Asia, equal to the global mean. Precipitation is likely to increase and extreme rainfall and other climatic events may occur more frequently. For India, extreme rains in South West monsoon and fewer rainy days along East Coast have been predicted.

Adaptive options like health-oriented management of the environment (including air and water quality, food safety, urban and housing design, and surface water management), and appropriate medical interventions are normally related to income levels. Therefore, poor income levels are more vulnerable. Human settlements are expected to be accordingly affected by changes in resource productivity or market demand for the goods and services causing structural and functional adaptations leading to migration, and resultant living conditions. Climate change would, therefore, result in increase in the absolute number of people at risk of hunger unless these outcomes can be countered through the development of cost effective technologies.

India has always been subject to a large degree of climate variability. This is likely to be accentuated by climate change. Warming trends along the west coast, in central India, the interior peninsula, and north-eastern India and cooling trends in north-west India and parts of south India have been observed. Coasts are projected to be exposed to increase in risk, including coastal erosion due to climate change and sea level rise. Geological Survey of India (GSI) projects that the glaciers of Himalaya are receding at varying rates in different regions. India is conscious of these challenges and has implemented several major programmes addressing the climate variability concerns. These include: (a) cyclone warning and protection, (b) coastal protection, (c) floods and drought control and relief, (d) major and minor irrigation projects, (e) control of malaria, (f) food security measures and (g) research on drought resistant crops.

The climate of India may be broadly described as tropical

monsoonal type. India's climate is affected by two seasonal winds: the North-East monsoon and the South-West monsoon. The North-East monsoon, commonly known as the winter monsoon blows from land to sea, whereas the South-West monsoon, known as the summer monsoon blows from sea to land after crossing the Indian Ocean, the Arabian Sea, and the Bay of Bengal. The South-West monsoon brings most of the rainfall during a year in the country.

India is a large developing country with nearly 700 million rural population directly dependent on climate-sensitive sectors (agriculture, forests and fisheries) and natural resources (water, biodiversity, mangroves, coastal zones, grasslands) for subsistence and livelihoods. Further, the adaptive capacity of dry land farmers, forest dwellers, fisher folk and nomadic shepherds is very low.

Climate change may alter the distribution and quality of India's natural resources and adversely affect the livelihoods of its people. With an economy closely linked to its natural resource base and climatically-sensitive sectors, India may face a major threat because of the projected change in climate. With climate change, there would be increasing scarcity of water, reduction in yields of forest biomass, and increased risk to human health. The contribution of India to the cumulative global emissions is only 5 percent. Thus, historically and at present, India's share in the carbon stock in the atmosphere is relatively miniscule when compared to its distribution over the nation's population.

6.6 National Mission for Sustainable Agriculture (NMSA)

NMSA seeks to address issues regarding sustainable agriculture in the context of risks associated with climate change by devising appropriate adaptation and mitigation strategies for ensuring food security, enhancing livelihood opportunities, and contributing to economic stability at the national level. Under NMSA, the adaptation and mitigation measures would be mainstreamed in the following areas:

- Research and development activities.
- Absorption of improved technology and best practices.
- Creation of physical and financial infrastructure and institutional framework.
- Facilitating access to information.
- Promoting capacity building.

While promotion of dry land agriculture would receive prime importance by way of developing suitable drought- and pest-resistant crop varieties and ensuring adequacy of institutional support, NMSA would also expand its coverage to rain-fed areas for integrating farming systems with livestock and fisheries so that agriculture continues to grow in a sustainable manner.

6.7 Prime Minister's Council on Climate Change

Government of India has set up an elaborate institutional mechanism to consider and address issues relating to climate change. A Council chaired by Prime Minister called Prime Minister's Council on Climate Change was constituted in June 2007 to coordinate national action for assessment, adaptation and mitigation of climate change. The Council provides the overall guidance to climate change related actions taken by various Ministries of the Government and other agencies.

Climate change needs to be accounted as one of the potential risk factors affecting the outcome of any development programme or process in the long-run. Specific institutional arrangements are needed within government to oversee and monitor development in the context of the above necessities.

Forecasting systems and adaptations in response to climate change will be the key concerns for future development. For living systems, study of the impact of factors considered subject to rapid changes and their adaptation potential against the changing factors will be important. For predictions and early warning systems, understanding of the processes which could show discernible changes well before the actual impact starts, is essential.

Vulnerabilities of India to climate change—in particular impacts on water resources, forests, coastal areas, agriculture, and health—should be identified. For this, co-ordinated efforts through the relevant scientific institutions, in both government and non-government sectors, are required. For specific thematic areas, nodal institutions can be identified and designated.

Initiatives are needed in the area of adaptation to climate change. The extension of action needed, obviously depends upon the expected success in mitigation of climate change and this is difficult to predict at this stage. However, since a substantial adverse change appears unavoidable at present, there is need to embark on a process of adaptation which would, at least, identify the initiatives needed to deal with those outcomes that are highly likely. This requires action in the area of agricultural research to evolve varieties that can cope with likely climate changes.

6.8 Conclusions

Intensive agriculture during the Green Revolution has brought significant land and water problems relating to soil degradation over exploitation of ground water and soil pollution due to the uses of high doses of fertilisers and pesticides. Understanding of environmental consequences of agricultural development helps to illustrate the nature and complexity of forces driving the productivity change and now these imply the concerns for food security of the region. Severe nutrient deficits of N, P and K occurred widely in harvested areas in both developing and least developed countries, particularly in the rice and wheat production systems in Asia, Central and South America, and Africa. Continuous depletion of soil N, P and K in most African countries and other least developed countries, coupled with low crop production levels, poses a real threat to agricultural sustainability and food security. Soil nutrient depletion caused by high production levels and decline in fertilizer use in recent decades in many developed countries is also a concern. Worldwide, soil fertility problems associated with the human-induced nutrient depletion are expected to continue.

An attempt should be made not only to increase the production but also to sustain the increased production without further degradation of the natural resources. Possible reclamation measures are required to be taken up for restoration of physical health of soils and its productivity. This may include integrated and sustainable monitoring and management of agriculture and forestry, which requires a focus on the issues such as collection and effective utilisation of land and water inventory data for land use planning, nutrient management, increased biomass productivity and need for diversification, re-enrichment of inherent fertility and agro-forestry, moisture conservation and water harvesting, the recharging of ground water reservoirs in areas of water table decline and agro-industrial watershed based planning.

7

Degradation of Soil Resource Base [1]

7.1 Problem of Land Degradation

Of all the human induced land degradation problems, the permanent loss of soil productivity due to erosion is the worst on a global scale. Accelerated erosion occurs nearly everywhere where agriculture is practiced and is irreversible in nature. Loss of nutrient rich fine soil not only reduces productivity, but also results in silting of water bodies and streams and induces release of soil carbon from particulate organic material, which contributes to global warming. Asia has suffered the most from human induced *soil erosion* than any other continent of which agricultural activities account for 25 percent (Wani et al., 2001). [2]

Lindert (2000) defines soil degradation more specifically as "any chemical, physical, or biological change in the soil's condition that lowers its agricultural productivity, defined as its contribution to the economic value of yields per unit of land area, holding other agricultural inputs the same". (Lindert notes that "[a] synonym for the soil's 'agricultural productivity' is soil's quality"). Examples of soil degradation include loss of top soil through erosion by water or wind, depletion of soil nutrients, loss of soil organic matter, compaction, waterlogging, salinization, and acidification. Soil degradation occurs as a result of both natural and human induced processes, such as agricultural production. This implies temporary and permanent weakening from a higher to lower status of productivity through unstable soil characteristics. The intensive cropping systems introduced by the Green Revolution impose much heavier demands on crop nutrients than traditional systems. Intensive agricultural practices can also result in changes in soil physical and chemical properties. To quote, *"through its impacts on agricultural productivity and environment, soil degradation leads to food shortage, which in turn leads to political and social instability. Thus, it affects the social fabric of the society"* (TERI, 1998; p. 61).

An inadequate and imbalanced nutrient use coupled with neglect of organic manures has resulted in multi-nutrient deficiencies in Indian soils. These deficiencies are becoming more critical for sulphur, zinc and boron. As the nutrient additions do not keep pace with nutrient removal by the crops, the fertility status of Indian soils has been declining fast under intensive agriculture and is now showing signs of fatigue, especially in the Indo-Gangetic plain.

The soils in India is having a net negative balance of about 8-10 million tonnes of NPK at the country level. The potassium (P) is the most mined nutrient from soils with the rate of its removal being 7 metric tonnes and in proportion to an addition of only one metric tonne. Sulphur deficiencies are also showing up in all parts of the country; being more rampant in the southern region.

7.2 Land Pollution and Land Degradation: Causes

Land, a non-renewable resource, is central to all primary production systems. Over the years, India's landmass has suffered from different types of degradations. Floods and droughts are recurrent phenomena in India, severely constraining agriculture. The problem of land degradation gets exacerbated due to large scale land erosion in which fertile top soil is lost.

As per the study conducted by National Bureau of Soil Survey and Land Use Planning, Nagpur, [a Regional Centre of Indian Council of Agricultural Research (ICAR)], an area of about 146.82 million hectares is subjected to various kinds of land degradation, viz. water and wind erosion, water logging, alkalinity/salinity, soil acidity and other complex problems across the country. Frequent occurrence of droughts, floods and climate change also have/had an impact on soil fertility and land degradation, thereby affecting foodgrains production in the country.

Land degradation occurring due to the natural and human induced causes like wind erosion and water logging, is one of the priority concerns in India. The varying degrees and types of degradation stem mainly from unsustainable use and inappropriate land management practices.

7.2.1 Intensive Agricultural Practices: Intensive agricultural practices that rely heavily on water, chemical fertilisers and pesticides have caused water-logging and salinity in many parts of the country. The expansion of the irrigation system without adequate steps for treatment of the catchment areas has exacerbated this. The quest for increased agricultural productivity has led to intensive cultivation of marginal lands causing their degradation.

7.2.2 Unsanitary Disposal of Solid Wastes: Pollution of land also results from the unsanitary disposal of solid wastes. In India, open dumping of such wastes of municipal and industrial origin on low lying land is a common phenomenon. This serves as a breeding ground for pests and disease carrying vectors.

7.2.3 Pressure of Human and Cattle Population: In addition to the wastelands, other areas such as deserts, drought-prone, flood-prone and tribal areas have been subjected to severe forms of degradation. The capacity of these lands is limited due to environmental factors. Pressures of human and livestock population have further compromised them.

7.2.4 Community Lands: Some of the most degraded lands in the country are the common property resources (CPRs). CPRs are resources on which people have an equal right of use. These resources include community pastures, community forests, wastelands and common dumping and threshing grounds. In spite of concerted efforts to check deforestation and the various afforestation schemes taken up during successive Five Year Plans, large tracks of forest continue to be classified as degraded.

7.2.5 Mining Activities: Land degradation is considered to be an unavoidable by-product of mining and has reached alarming proportions, mainly due to over-exploitation and mismanagement of natural resources. Mining activity often leads to environmental problems like land degradation, particularly in open cast mining and land subsidence in underground mining. Open cast mining in areas with forest cover causes deforestation.

7.2.6 Industrial Effluents: Soil pollution from heavy metals due to improper disposal of industrial effluents, along with the excessive use of pesticides and mismanagement of domestic and municipal wastes, is becoming a major concern. Though no reliable estimates are available to depict the exact extent and degree of this type of land degradation, it is believed that the problem is extensive and its effects are significant. Some commercial fertilizers also contain appreciable quantities of heavy metals, which have undesirable effects on the environment. The indiscriminate use of agrochemicals, such as fertilizers and pesticides, is often responsible for land degradation.

7.2.7 Over Extraction from Forests: Loss of vegetation occurs as a result of deforestation, cutting beyond the silviculturally permissible limits, unsustainable fuel wood and fodder extraction, shifting cultivation, encroachment into forest lands, forest fires and overgrazing, all of which subject the land to degradational forces.

7.2.8 Other Causes: Other important factors responsible for large scale degradation are the following:

1. Non-adoption of adequate soil conservation measures.
2. Improper crop rotation.
3. Indiscriminate use of agro-chemicals such as fertilizers and pesticides.
4. Improper planning and management of irrigation systems.
5. Extraction of groundwater in excess of the recharge capacity.

6. Extension of cultivation to lands of low potential or high natural hazards.

In addition, there are a few underlying or indirect pressures such as land shortage, short-term or insecure land tenancy, open access resource, economic status and poverty of the agriculture dependent people, which are also instrumental, to a significant extent, for the degradation of the land.

7.3 Consequences of Land Degradation

Land degradation has a direct bearing on the productivity of soil, its vulnerability to rainfall variations, scarcity of drinking water, fodder and fuel wood. Given the inter-linkages of crop production and livestock economy and environment, land degradation has a major impact on the livelihoods of the people, especially in rural areas.

The ever increasing generation of solid waste is a growing problem in all cities. Uncollected garbage is not only visually ugly, it also causes diseases. Dumping garbage into landfills is not an attractive option in land scarce cities and should be minimised. Most industrial countries now require their citizens to segregate wastes at home into recyclable products. Cities and towns must be encouraged to evolve systems with citizen participation, for segregation of waste at point of origin, maximising recycling and safe disposal of the rest. India's villages also have a huge problem of garbage disposal which must be addressed through self-help groups (SHGs) and panchayti raj institutions (PRIs).

Direct consequences of agricultural development on the environment arise from intensive farming activities, which contribute to soil erosion, land salination and loss of nutrients. The introduction of *green revolution* in the country has been accompanied by over-exploitation of land and water resources and excessive usage of fertilizers and pesticides. Shifting cultivation has also been a major factor responsible for land degradation in hilly areas. Leaching due to extensive use of pesticides and fertilizers is a major source of contamination of water bodies.

The extent of agricultural intensification and extensification is characterized by an increase in cropping and irrigation intensity and the unbalanced use of chemical fertilizers, pesticides and insecticides. It has also led to land degradation, overexploitation of underground water resources and increased use of chemical fertilizers, leading to eutrophication and water pollution in some regions.

Enhanced intensification and extensification also leads to salination, alkalization and water logging in irrigated areas, along with eutrophication of water bodies and ill health of oceans, leading to loss of biodiversity. For achieving and maintaining food security and sustainable forestry, controlling of land/soil erosion is extremely vital. It is essential to control soil erosion in order to attain and maintain food security, sustainable forestry and agricultural and rural development.

7.4 State-wise Soil Fertility Status

The deficiencies could be assumed to be occurring in 40-45 percent of districts covering about 60 million ha of net sown area. State-wise soil fertility status and extent of micro nutrient deficiencies in Indian soils is given in Tables 7.1 and 7.2.

Zinc deficiency is recognized as a nutritional disorder in rice on sodic and calcareous soils that have a high pH. The iron deficiency has been noticed in rice on coarse-textured soils. Similarly, the deficiency of manganese is increasingly appearing in wheat when it is grown after rice, on coarse textured alkaline soil having low organic matter content, high available phosphorus or silicon and high manganese-calcium ratio. Boron deficiency is becoming more widespread in red and lateritic, acidic, coarse textured alluvial soil. The limiting nutrients do not allow for the full expression of other nutrients, thereby, lowering the fertilizer responses and crop productivity. Soil-test based balanced and integrated nutrient management through conjunctive use of both inorganic and organic sources of plant nutrients is the ideal solution to prevent such deterioration of soil health.

Table 7.1: State-wise Soil Fertility Status

Name of the State	Category of Fertility Status		
	Nitrogen	Phosphorus	Potassium
Andaman & Nicobar	Low	Low	Low
Andhra Pradesh	Low	Low	High
Arunachal Pradesh	High	Low	High
Assam	Medium	Medium	Medium
Goa	High	Medium	High
Gujarat	Medium	Low	High
Haryana	Low	Low	Medium
Himachal Pradesh	High	Medium	Low
Jammu & Kashmir	Medium	Low	Low
Karnataka	Medium	Medium	High
Kerala	Low	Medium	Medium
Madhya Pradesh	Medium	Medium	High
Maharashtra	Low	Low	High
Manipur	Low	Low	Medium
Meghalaya	High	Low	Medium
Mizoram	High	Low	Medium
Nagaland	High	Medium	Low
Odisha	Low	Low	Medium
Pondicherry	Low	Low	Medium
Punjab	Medium	Medium	High
Tamil Nadu	Low	Medium	High
Tripura	Medium	Medium	Medium
Uttar Pradesh	Low	Low	Medium
West Bengal	Medium	Medium	Medium
All India	Low	Medium	High

Source: Motsara, M.R. (2002).

In addition, split application, placement of fertilizer, use of slow releasing N-fertilizers and nitrification inhibitors and inclusion of leguminous crop in crop rotation are also advocated.

A considerable amount of information on nutrient uptake and removal by crops and cropping system is now available. In most cases, different balance sheets are not comparable due to vastly different assumptions and computation methodologies. Several aspects of nutrient uptake, removal, and balances have been dealt with in detail elsewhere (Kanwar and Katyal, 1997; Tandon, 2004).

Table 7.2: Extent of Micronutrient Deficiency in Indian Soils

State/Union Territory	Samples	Percent Samples Deficient					
		Z	C	I	Ma	B	Mo
Andhra Pradesh	8,158	49	<1	3	1	-	-
Assam	12,166	34	<1	2	20	-	-
Bihar	19,214	54	3	6	2	38	-
Delhi	201	20	-	-	-	-	-
Gujarat	30,152	24	4	8	4	2	10
Haryana	21,648	60	2	20	4	0	28
Himachal Pradesh	155	42	0	27	5	-	-
Jammu & Kashmir	93	12	-	-	-	-	-
Karnataka	27,860	73	5	35	17	32	-
Kerala	650	34	3	<1	0	-	-
Madhya Pradesh	32,867	44	<1	7	1	22	18
Maharashtra	515	86	0	24	0	-	-
Meghalaya	95	57	2	0	23	-	-
Odisha	16,040	54	-	0	0	-	-
Pondicherry	4,108	8	4	2	3	-	-
Punjab	16,483	48	1	14	2	13	-
Rajasthan	183	21	-	-	-	-	-
Tamil Nadu	28,087	58	6	17	6	21	-
Uttar Pradesh	26,126	46	1	6	3	24	-
West Bengal	6,547	36	0	0	3	68	-
All	251,547	48	3	12	5	33	13

Z = Zinc; C = Copper; I = Iron; Ma = Manganese; B = Boron and Mo = Molybdenum

Source: Project Report, Coordinated Micronutrient Research Project, Indian Institute of Soil Science (IISS), Bhopal.

7.5 Soil Erosion and Productivity Loss

The severe data limitations and differences in methods have resulted a wide range of estimates of past on potential impacts of soil degradation on agricultural productivity and production at various scales. Various studies pointed out the loss in productivity is directly linked to the severity of soil erosion. Den Biggelaar et al. (2004) analyzed six crops—maize, wheat, soybeans, sorghum, millet and potatoes—that together accounted for three-quarters of the experiments conducted. Mean yields and yield losses per ton of soil erosion were calculated across soils for each crop and region (Table 7.3). It is noted that yield losses may accelerate, remain constant, or decelerate as soil erodes, depending on soil type and crops.

Table 7.3: Mean Loss in Annual Yield per Ton of Soil Erosion

Continent	Crop	Estimated Mean Production/Value Loss (percent per year)
Africa	Maize	0.49
Asia	Maize	0.59
	Millet	0.51
	Soybeans	21.08
	Wheat	0.29
Australia	Potatoes	0.12
	Wheat	0.67
Europe	Millet	0.23
	Potatoes	0.04
	Soybeans	0.22
	Wheat	0.04
North America	Maize	0.15
	Potatoes	3.98
	Sorghum	0.06
	Soybeans	0.24
	Wheat	0.11
Central South America	Maize	0.94
	Potatoes	0.01
	Soybeans	0.33
	Wheat	0.27
Global Total	Maize	0.42
	Potatoes	0.60
	Millet	0.48
	Sorghum	0.06
	Soybeans	0.08
	Wheat	0.20
	Total	0.30

Source: Den Biggelaar et al. (pp. 87-88) (2004); Production data from FAO Stat, 2000.

Several studies of productivity impacts have been conducted at a global scale. Controlling for agricultural inputs and institutional changes, Huang and Rozelle (1995) found that environmental degradation, primarily in the form of erosion and salinization, reduced grain yields in China by about 5 percent between 1976 and 1989 (about 0.4 percent per year). They noted that degradation's impacts vary by crop and region; losses to erosion were especially high in northern China. Lindert (2000) noted concerns about the accuracy of statistics, perceptions, and analysis of cultivated land and land quality in China. In an econometric analysis of the interaction between crop production and soil quality parameters between 1930s and the 1980s, he found that agricultural intensification in China has depleted nitrogen and organic matter in some cases but increased soil endowments of phosphorus and potassium, while concerns about soil erosion have been greatly exaggerated. Losses in nitrogen and organic matter have had no clear effect on crop yields in China because of the ability of commercial fertilizer to compensate for those properties. Lindert found similar results for Indonesia.

Soil nutrient depletion and imbalance are also widespread problems in Asia. Dobermann et al. (1995) computed nutrient balances for rice-based farming systems in several eastern Asian countries, and reported that the K deficit ranged from 25 kg/ha/yr to 70 kg/ha/yr. Lin et al. (1996) estimated nutrient balances on 71 rice-paddy farms in south China and found a surplus of N and P but a deficit of K. Mutert (1996) reported negative balances of N, P and K in Bangladesh, Indonesia, Myanmar, the Philippines, Thailand, and Vietnam. In contrast, an oversupply of nutrients was observed in Japan, the Republic of Korea and Malaysia. Therefore, Mutert (1996) concluded that in Asia, negative nutrient balances for major crops and rice occurred in lower-income countries with large and growing populations, whilst an oversupply and positive balance occurred in high-income countries with small and stable populations.

Pagiola (1995) notes that rice yields in Bangladesh are stagnant or declining despite rising input use, strongly suggesting that yields are being reduced by land degradation in the form of nutrient imbalances and other sub-optimal soil properties. Pagiola (1998) argues that erosion problems in El Salvador have been exaggerated. Perhaps one-third of fields on moderate slopes and two-thirds of those on steep slopes experience productivity problems due to erosion.

Focusing on the United States, Pimentel et al. (1995) used an empirical model (not described) to estimate losses of water, organic matter, available nitrogen, and other properties associated with soil erosion at a rate of 17 tons per hectare per year (characteristic of U.S. crop land in the early 1980s). These losses were in turn associated with a decline in crop productivity (maize yields) of 8 percent per year.

Sehgal and Abrol (1994) presented estimates of the severity of erosion in India. They cited experimental data on erosion-productivity relationships based on three-year experiments on vertisols in Nagpur, Maharashtra, which showed very small losses if erosion is kept to less than 10 tons/ha/year, but very high losses under higher rates of erosion. They found that erosion at the rate of 10-20 tons/ha/year causes yields to fall by 7.8-34.3 percent per year, and erosion at the rate of 20-40 tons/ha/year, causes losses of 58-68 percent per year. These findings suggest that for most crops, erosion at rates of more than 10 tons per year would lead to practically zero yield in less than two decades.

7.6 Environmental Implications of the Green Revolution

To analyse the environmental implications of the Green Revolution in the state of Punjab, Brar (1999) constructed a Green Revolution intensity indicator for each of the 118 "development blocks" in the State. A simple cross tabulation of some of the measures and indicators of externalities associated with Green Revolution technology, namely decline in water tables, soil nitrogen deficiency and soil salinity, with the intensity index, indicated no direct relationship between the Green Revolution intensity and severity of negative externalities (Brar, 1999). For example, only one-third of the blocks in high intensity (more than 0.75) region of Punjab recorded a fall in water table by more than 3 meters as against a majority of blocks in the 0.25 to 0.50 intensity blocks, which showed a decline in water table by more than 3 meters. This shows that local conditions of sub-surface water have their own role to play in the dynamics of changing water tables. Similarly, some of the badly affected soil salinity areas were found to be in Bathinda and Faridkot districts which fell under the intensity index of less than 0.25 and 0.50, respectively.

Ali and Byerlee (2001) argued that a positive trend in total factor productivity (TFP) is inadequate as an indicator of sustainable production growth because the effects of resource degradation may be masked by improvements in technology. Using district-level data on irrigated agriculture in Pakistan's Punjab province for 1971-94, they found that both land and labour productivity grew by about 2.5 percent annually over the period, the former due to the introduction of Green Revolution technologies (such as improved seeds, fertilizer, and water control) and the latter due to subsequent mechanization. TFP growth in the province averaged 1.3 percent per year for crops as a whole but declined in the rice-wheat system. Resource degradation (in the form of depletion of soil organic matter and available phosphorus, increased soluble salts and pH, and reduced water quality) was found to have lowered TFP by 58 percent on an average and to have more than cancelled the positive effects of technological change in the rice-wheat system.

Productivity losses are difficult to quantify due to a lack of data, making the conclusion seems inescapable, as the evidence is insufficient to clearly link productivity declines to decreased soil fertility because most farmers (especially in technologically advanced areas) have applied fertilizers nutrients like nitrogen and phosphorus at close to recommended levels (Hobbs and Morris, 1995). Deficiencies in micro nutrients like zinc, boron, and manganese are also cited as a yield limiting factor but yield experiments with these nutrients rarely give conclusive results (Hobbs and Morris, 1995). Organic matter also plays a role in balancing soil chemistry although it is not yet fully understood. In major producing areas, soil acidification and salinization are also major issues that may have contributed to past yield declines (Fischer, 1994).

Nutrient balance sheets in most soils of India have been deficient and continue to be so (Tandon, 2004). In June 2012, Indian Institute of Soil Science (IISS) uploaded maps of 18 states that showed nutrient deficiency district by district. The maps paint a scary picture. Large parts of India are deficient in two or more critical nutrients. Areas like the Indo-Gangetic plains—Punjab, Haryana, Uttar Pradesh and Bihar—which produce nearly 50 percent of our grains and feed about 40 percent of our population—are seeing multiple deficiencies. This is primarily because nutrient removals by crops far exceed the nutrient additions through manures and fertilisers. For the past 50 years, the gap between removals and additions has been estimated at 8mt to 10 mt of $N+P_2O_5+K_2O$ per year (Tandon, 2004). This has been the case in the past, at present, and is likely continue into the future. To this extent, the soils are becoming depleted—the situation is akin to mining soils of their nutrient capital, leading to a steady reduction in soil nutrient supplying capacity.

On top of this deficit are the nutrient losses through various other means. For example, nutrient losses through soil erosion are alarmingly large, but are rarely taken into account. Nutrient loss through soil erosion is second only to nutrient removal as a result of crop production. An annual loss of 8 mt plant nutrients has been mentioned through 5.3 billion tonnes of soil lost by water erosion (Prasad and Biswas, 2000). Estimates of removals through leaching and gaseous losses are not available.

Estimated rates of soil erosion and its consequences also vary. Also, the impact on productivity of a given rate of soil loss depends greatly on the soil type, its original depth, and the time period over which the rate of erosion is sustained. Aggregate figures on soil loss per hectare can be misleading, because often they are estimated on the basis of soil loss measurements on individual experimental plots. Often under this approach, all soil that erodes in a given area is assumed to disappear, to be lost permanently. Observations on farmers' fields in India, however, reveal that sometimes erosion can be a loss to one farmer but a gain to another on whose land the eroded soil is deposited. Many farmers actually encourage erosion on one part of a plot in order to concentrate soil where it can contribute to greater overall productivity (Kerr and Sanghi, 1992). What this means is that a given estimated rate of soil loss for a given region probably is accurate on some plots or parts of some plots, but other areas are probably losing a smaller amount and some are probably even gaining more than they are losing. With this caveat in mind, we present some estimated rates of soil loss for India.

Singh et al. (1992) produced an iso-erosion map of India that delineates areas by their rate of soil loss per ha. On the high end, ravine areas (mentioned above) are estimated to lose over 40 tons/ha per year, and the Shivalik Hills at the foot of the Himalayas are thought to lose over 80 tons/ha per year. Rates in the Indo-Gangetic Plains are estimated at uniformly less.

The adverse impacts of irrigation induced soil salinity and water-logging are evident at the farm level in the form of decline in crop productivity, and decline in input use levels (Joshi, 1987; Chopra, 1989; Singh and Singh, 1995; Singh et al., 2002). Joshi (1987) and Chopra (1989) estimated losses due to soil salinity using the Cobb-Douglas production function and dummy variable techniques. At the micro-level Singh and Singh (1995), using the Cobb-Douglas Production function for a sample of 248 respondents of Western Jamuna and Bhakra Canal in Haryana, observed a huge cost in non-land resource use on problem soils (salt affected) as compared to normal soils. This consequently resulted in low farm production and income. Employment opportunities were also greatly reduced on problem soils.

Erosion rates in the black soil region (vertisols) of the country, occupying a 64 million ha area in Karnataka, Andhra Pradesh, Madhya Pradesh and Maharashtra states were around 20 tons/ha/yr (Singh et al., 1992). However, this estimate was quoted much higher by Dhruva Narayana and Ram Babu (1983) at 64 tons/ha/yr while runoff was estimated as high as 40 percent. On an average, about 24 kg N/ha is lost per year (Kanwar et al., 1989). Krantz et al. (1978), argued that soil erosion due to rainy season fallowing is the reason why the proportion of vertisols area classified as deep in Sholapur district of India has shrunk from 46 percent to only 29 percent in a span of 75 years.

Tan et al. (2004) observed that in both developing and least developed countries, the rate of increase in crop yield has decreased with increasing fertilizer use since the 1960s.

The greatest decline in the yield/NPK fertilizer ratio was observed in least developed countries, which was a result of low input into and high depletion of nutrients out from soils. Low response can be attributed to two reasons: the limited yield potential of a crop, and the effect of the most limited nutrient element.

Obviously, the yield potential of a crop should not become a limit to these four crops because their actual yield levels in both developing and least developed countries were much lower than those in developed countries.

We also observed the same for the data used in the current analysis. The quadratic polynomial regression has been fitted between fertiliser use per hectare and yield per hectare to observe the responsiveness behaviour. The sigmoid or S-shaped curve, as observed in cases of rice and wheat, presents an alarming picture of diminishing returns of fertiliser use in case of yield in high productive states. There seems to be democratization in the ratio of yield to fertilizer use among high and low rice and wheat productive states in the last two decades. This clearly shows the wheat and rice production in India is operating at the top of the curve with diminishing returns to the scale following increasing use of the fertilisers.

Rice and wheat are heavy users of nutrients. Nutrient imbalance and mining by these cereal crops had led to nutrient deficiencies and poor soil quality in some areas. The rice-wheat systems are enormously complex, with numerous productivity and sustainability problems. Some of these problems are late sowing, low water and nutrient use efficiency, groundwater depletion (in some areas) or waterlogging (in other areas), salinity and sodicity (in specific areas), and build-up of weeds, pests, and diseases. These problems are closely linked through system interactions; e.g., late sowing.

Reduced and zero tillage improve timelines of sowing and through the same system interactions help farmers cope with crop production problems. These practices improve nutrient use efficiency and reduce weed germination.

7.7 Pest Problems

Chemical inputs used to increase agricultural productivity, such as inorganic fertilizers and pesticides have been associated with many negative direct and indirect human health impacts. Pesticides as such are toxic chemicals and represent risks to users. In developing countries, where users are often illiterate, ill-trained and do not possess appropriate protective equipment, the risks are magnified. Furthermore, comprehensive bodies of legislation to regulate the use and distribution of pesticides often do not yet exist. It is estimated that only 0.1 percent of applied pesticides reach the target pests, leaving the bulk of the pesticides (99.9 percent) to impact the environment (Pimental, 1995). Human pesticide poisonings and illnesses are clearly the largest "environmental costs" paid by a society for pesticide use.

The study by Rolaand Pingali (1993) found that:

- Pesticide use has a significant positive association with the incidence of multiple health impairments in Philippine rice farmers, even after accounting for other effects (e.g. age, smoking, drinking habits and nutritional status).
- The average health cost for farmers exposed to pesticides was approximately 40 percent higher than that for the unexposed farmers. Even after accounting for other factors, health costs increase by 0.5 percent for every 1 percent increase in insecticide dose above the average level.
- Health impairment leads to a loss in labour productivity.

Insect pests, diseases and weeds inflict enormous losses to the potential agricultural production. For instance, insect pests, diseases and weeds cause considerable damage to potential agricultural production. Evidences indicate that pests cause 25 percent loss in rice, 5-10 percent in wheat, 30 percent in pulses, 35 percent in oilseeds, 20 percent in sugarcane and 50 percent in cotton (Dhaliwal and Arora, 1996).

Until recently, chemical pesticides were increasingly relied upon to limit the production losses. Pesticide use in India increased from a mere 15 g/ha of gross cropped in 1955-56 to 90 g/ha in 1965-66. Introduction of green revolution technologies in mid-1960s gave a fillip to pesticide use, and in 1975-76, it had increased to 266 g/ha, and reached a peak of 404 g/ha in 1990-91 (Birthal, 2003). About half of the total pesticides used in agriculture go towards controlling insect pests and diseases of cotton, which occupies only 5 percent of the total cultivated area. Cotton receives as many as 15-20 rounds of insecticide sprays right from the vegetative stage till its maturity. According to the estimates by Birthal and Jha (1997), one hectare of cotton receives 3.75 kg. of pesticides. Rice, with an area share of 24 percent, accounts for 17 percent of the total pesticide use. Although, there is a paucity of reliable time-series information on pest-induced production losses, anecdotal evidences suggest increase in losses

(Dhaliwal and Arora, 1996), despite increase in the pesticide use. The paradox is explained in terms of rising pest problem, technological failure of chemical pesticides and changes in production systems. Nevertheless, pesticide use has started declining since 1990-91, reaching 265 g/ha in 1998-99, without affecting the agricultural productivity (Birthal, 2003).

7.8 Strategies to Check Land Degradation

The prevention of land degradation and the augmentation of the carrying capacity of land to provide food, fuel and fodder requirements have, therefore, been a primary concern of the Government. Special Area Development Programmes have been funded to mitigate the harsh living conditions of people through better land management, water harvesting and conservation practices on a watershed basis.

India has paid a good deal of attention to harness its water resources by way of construction of major, medium and minor irrigation projects and the development of ground water resources. Organizations have also been set up in this field, namely Central and State Irrigation Departments, Central Water Commission and the Central Ground Water Board. However, very little attention has been paid to the proper management of land and soil resources with the result that they have suffered very serious degradation. The losses which the country is bearing on account of the continuing degradation of its land resources are of staggering dimensions and constitute one of the important threats to India's economic progress.

Land degradation problem could be tackled to an extent by suitable policies that would internalize the issue into proper decision-making. At the macro level, the existing database on land use statistics cannot adequately facilitate the analysis of land degradation and its impact. Changes in the classification of land use statistics are needed in order to study its impact. Advanced technology like remote sensing could go a long way in generating vital information on different dimensions of land degradation.

The information base on which farmers make decisions is

incomplete in terms of internalizing rapid changes in soil and water quality variables. Hence, there is a need to move towards more sustainable practices such as integrated pest management and land-conserving crop rotations. Research needs to be focused on measures such as integrated crop management. An integrated approach to the problem of degradation, linking agriculture and environment, is yet to be attempted even at the policy level.

Overuse of resources has been identified as a major cause of land degradation. Creation of forest/biomass resources in all the cultivable vacant lands should be taken up for strengthening life support system of communities and maintaining soil/water regimes.

7.9 Programmes/Schemes for Treatment of Degraded Lands

With a view to prevent land degradation and to sustain agricultural production, Department of Agriculture and Cooperation of the Ministry of Agriculture, Government of India is implementing various soil and water conservation programmes for treatment/reclamation of degraded lands. The objectives of various schemes/programmes are as under:

1. Implementation of soil and water conservation programmes for the prevention of soil erosion and land degradation in order to sustain agriculture production on watershed approach basis.
2. Formulation of policies and strategies to address the issues regarding sustainable agriculture in the context of risks associated with climate change and suggesting appropriate adaptation and mitigation measures to combat adverse effect of climate change on agriculture production including natural resources.
3. Reclamation and development of land affected by alkalinity and acidity.
4. Providing scientific database on soil and land uses for watershed planning and programme implementation.
5. Prevention of runoff and premature siltation of the

reservoirs, thereby sustaining irrigation potential and water storage capacity of the various reservoirs across the country.

6. Encouraging settled cultivation in lieu of shifting cultivation among the Jhumia families of the States of North East Region and restoration of ecology and checking land degradation.
7. Creation of employment opportunities and improving the socio-economic conditions of the rural community.
8. Development of human resources through training and capacity building programmes of state officials working in the area of soil and water conservation.

The Central sector schemes and Centrally sponsored programmes being implemented by the Government are described below.

7.9.1 Soil and Land Use Survey of India (SLUSI): SLUSI is a subordinate office of the Department of Agriculture and Cooperation, established in the year 1958 with its headquarters in New Delhi. SLUSI has 7 Regional Centres located at NOIDA (Uttar Pradesh), Kolkata, Bengaluru, Nagpur, Hyderabad, Ahmedabad and Ranchi. The major mandate of SLUSI is to conduct various types of soil and land resource surveys and provide a soil-related database for the planning and implementation of programmes relating to soil and water conservation for the development of degraded lands. The main activities of SLUSI are the following:

1. Rapid reconnaissance survey (RRS) of the catchments of river valley projects and flood prone rivers for categorization of watersheds.
2. Detailed soil survey (DSS) of selected *very high* and *high* priority watersheds to provide a sound database for execution of soil conservation measures as well as for scientific land use planning.
3. Soil resource mapping (SRM).
4. Development of GIS-based web server.
5. Organizing short training courses on soil surveys and data

use for planning of watershed management.

6. Consultancy projects in soil mapping.

The organization is engaged in conducting soil surveys to provide database for land-based developmental programmes for the user organizations. Besides, SLUSI provides training to the State officials drawn from the state departments such as agriculture, soil conservation, soil survey, forestry and irrigation.

7.9.2 Soil Conservation Training Centre, Damodar Valley Corporation (DVC), Hazaribagh, Jharkhand: This Centre organizes medium and short duration training courses every year for field functionaries and project officers of the State Governments engaged in implementation of soil and water conservation programmes.

7.9.3 Soil Conservation in the Catchments of River Valley Projects and Flood Prone Rivers: Presently, this Centrally-sponsored programme is being implemented in 60 catchments of 27 States (all the States except Goa) including all the States of North Eastern Region of the country. The total catchment area of ongoing programme is about 113.51 million hectares, out of which an area about 30.15 million hectares is categorized as a priority area, needing urgent treatment under this programme. The catchments are mostly inter-state in nature and all categories of lands viz. agriculture, waste lands and forests are taken up for treatment in an integrated manner on watershed approach basis. The main objectives are the following:

1. Prevention of land degradation by adopting a multi-disciplinary integrated approach for soil conservation and watershed management in catchment areas.
2. Improvement of land capability and moisture regime in the watersheds.
3. Promotion of land use to match land capability.
4. Prevention of runoff and premature siltation of the reservoirs, thereby sustaining irrigation potential and water storage capacity of the various reservoirs across the

country.

Under this programme, various soil and water conservation measures are implemented on watershed approach. With a view to assess the impact of soil and water conservation measures, a system of continuous monitoring of rainfall, runoff and sediment parameters prior to, during and after treatment of the watershed are collected and analyzed for assessing the impact of watershed interventions. For this purpose, an amount of 3 percent of the allocated funds are earmarked for meeting the cost of establishment of sediment monitoring stations (SMS) and also to meet the operational cost of these SMSs.

7.9.4 Reclamation and Development of Alkali and Acid Soils (RADAS): This Programme was restructured and launched during the Eleventh Five Year Plan (2007-12) for development of alkali and acid soils in the states of Arunachal Pradesh, Mizoram, Gujarat, Haryana, Punjab, Karnataka and Rajasthan.

The Programme aims at improving physical conditions and the productivity status of alkali and acid soils for restoring optimum crop production. The major components permissible under the programme are on farm development, viz. land levelling, bunding, community drainage systems, application of soil amendments, organic manures, crop/horticultural/fuel wood production etc.

7.9.5 Watershed Development Project in Shifting Cultivation Areas: This Scheme was launched during the Eighth Five Year Plan (1992-1997) in 7 states of North Eastern Region, namely Arunachal Pradesh, Assam, Manipur, Meghalaya, Mizoram, Nagaland and Tripura with 100 percent Special Central Assistance to the State Plan. The major objectives of the scheme are the following:

1. Protection of hill slopes of Jhum areas through soil and water conservation measures on a watershed basis to reduce further land degradation.
2. Encouragement and assistance to Jhumia families to

develop Jhum land for productive uses with improved cultivation and suitable package of practices leading to settled cultivation.

3. Improvement of the socio-economic status of Jhumia families through household/land-based activities.
4. Mitigation of the ill effects of shifting cultivation by introducing appropriate land use methods through improved technologies.

Under this scheme, treatment of arable and non-arable land through drainage line treatment, water harvesting structures, farm ponds, horticulture, afforestation, silvipasture, crop demonstration etc. are taken up.

7.9.6 National Mission for Sustainable Agriculture (NMSA): NMSA is one of the 8 missions under the National Action Plan on Climate Change (NAPCC). It seeks to address issues regarding sustainable agriculture in the context of risks associated with climate change by devising appropriate adaptation and mitigation strategies for ensuring food security, enhancing livelihood opportunities and contributing to economical stability at the national level.

NMSA identifies 10 key dimensions for promoting sustainable agricultural practices, which will be realized by implementing a Programme of Action that covers both adaptation and mitigation measures through 4 functional areas, namely research and development, technologies, products and practices, infrastructure and capacity building. While recognizing the role of modern technologies and research in promoting the sustainability of agricultural production, NMSA also emphasizes the need to harness traditional knowledge and agricultural heritage for insitu conservation of genetic resources.

7.10 Fly Ash Utilization

Fly ash is the finely divided mineral residue resulting from the combustion of ground powdered coal in electric generating plant. Fly ash consists of inorganic matter present in the coal that has been fused during coal combustion. The fast increasing demand of power coupled with its dependence on coal for at least 2/3rd of its energy requirement is generating large volume of fly ash. Generation of about 40 million tonnes of fly ash during 1994 increased to about 100 million tonne in 2001 and 130 million tonne in 2007. The projections made by Planning Commission as well as Ministry of Power up to 2031-32 indicate that 2/3rd of power generation in the country would continue to depend on coal. The annual generation of fly ash is expected to be around 175 million tonne by end of Eleventh Five Year Plan (2007-12), 225 million tonne by end of Twelfth Five Year Plan (2012-17) and around 500 million tonne by 2031-32.

The first Fly Ash Notification was issued by Ministry of Environment and Forests (MoEF) in September 1999 to regulate the disposal of fly ash and ensure its proper utilization. Restriction was imposed to the extent that all brick kilns within the radius of 50 kilometres from coal/lignite based thermal power plants should use 25 percent fly ash while making the bricks.

A second notification making amendments was issued in August, 2003, increasing the radius from the thermal power plants to 100 kilometres.

The implementation of this notification since 1999 resulted in steady increase in the utilization of fly ash. However, the utilization has not reached 100 percent. Therefore, certain additional measures were required to further promote and facilitate its use to meet the following objectives:

1. To protect environment.
2. To conserve top soil.
3. To prevent dumping of fly ash from thermal power stations on land.
4. To promote utilization of ash in the manufacture of building materials and construction activity.

MoEF issued another notification dated November 3, 2009 to amend the existing fly ash notification of 1999 as amended in the year 2003. The main amendments proposed in the notification include the following:

1. Clay brick manufacturers have been excluded from the purview of this notification, as they are usually in the unorganized sector and the utilization of fly ash by them is not significant. The construction agencies engaged in the construction of buildings have been made responsible for use of only fly ash based products for construction.
2. Minimum fly ash content for building material/products to qualify as *fly ash based products* category has been stipulated.
3. Use of certain minimum percentage of fly ash for stowing of underground mines and backfilling of opencast mines has been made mandatory.
4. All coal/lignite based thermal power plants have been allowed to sell fly ash to the user agencies except the pond ash and mound ash which should be made available free of cost.
5. In addition, at least 20 percent of dry fly ash should also be made available free of charge to the fly ash or clay fly ash bricks, tiles and blocks manufacturers on priority basis.
6. Fly ash utilization targets for thermal power plants/expansion of plants, commissioned before and after issue of this notification, have been stipulated.
7. A monitoring committee to monitor the implementation of the provisions of the notification and submit its recommendations/observations every 6 months has been proposed.
8. All the financial institutions and agencies which fund construction activities shall include a clause in their loan/grant document for compliance of the provisions of this notification.

7.11 Mining and Land Use Management

Mineral development is one of a number of competing land uses. Due to lack of planning and other frameworks to balance and manage possible uses there are problems and disagreements in the matter of control, use, and management of land. Exploration and mining pose significant challenges in terms of land access and management. The most appropriate use of land is best decided within an integrated land planning framework that balances competing interests such as between national and local levels or between mining and conservation. There are trade-offs that may be made in order to generate benefits in one domain, but decisions on these should not be taken without inclusion of and negotiation with all those likely to be directly affected. A broad consensus must be achieved on the management of protected areas (e.g. forests) and the trade-offs involved. The local stake in the success of protected areas and resources available to manage them needs to be ensured.

The decision whether or not to mine a certain area should be undertaken through a democratic decision-making process and should be based on an integrated assessment of ecological, environmental, economic, and social impacts. The planning process will be more effective in the presence of equitable and inclusive rules of tenure, compensation schemes for those affected, and strong governance, including mechanisms for arbitration where necessary.

To sum up, it is necessary to ascertain the extent to which land now lying waste can become available for cultivation. Increasing population also means withdrawal of areas now under farms for building houses. Development of communications such as roads, railways, and airways may take up fertile land. Owing to rapid urbanisation and growth of large cities, land is needed for parks and open spaces. Irrigation dams may submerge fertile areas. Industrial plants and other establishments also require substantial areas. In the context of all these developments, wherever fertile land can be saved efforts should be made to do so. This indicates the need for a comprehensive inventory of land and for greater refinement in land classification and continuous attention to land use.

It is important to appreciate the inextricably close relationship between land and water management. Water, which is a renewable resource, can in fact be put to good use only if the land on which it falls, and the land to which it is applied, are properly cared for.

Land, which is for all purposes a non-renewable and inelastic resource, must be managed in such a manner as to be benefited rather than suffer damage as a result of its contact with water. The key to environmental quality, therefore, lies in scientific land and water management.

Endnotes

1. I am grateful to T.V. Jayan of Centre for Science and Environment (CSE) for some useful discussion on the soil degradation, which helped me to prepare this section. Jayan had written a cover story on the soil degradation entitled, *"The Last Action Heroing: Indian Soil, Up Against Heavy Odds"* in science and environment fortnightly, *Down to Earth*, January 31, 2003.
2. It has been estimated that globally among the causes of soil degradation, overgrazing accounts for 34 percent followed by deforestation 30 percent, agricultural activities 28 percent, overexploitation and industrialization are responsible for 7 and 1 percent of the degraded land, respectively (UNEP, 1993).

8

Degradation of Water Resources

8.1 Importance of Water

Without water all life ceases. Water has defined life in the Indian subcontinent for thousands of years. Water is used for drinking, domestic and municipal uses, agricultural irrigation, industries, navigation, recreation etc. The used water becomes contaminated and is called waste water which contains residuals of the activities taking place in each of them. Water resources are unevenly distributed over time and space.

By and large, rainfall, surface and ground water resources have been sufficient in providing adequate water to the Indian population. Rise in demand and development pressures are changing the characteristics of water in India. Erosion in the watershed due to the fast growing development and poor land management practices is increasing siltation and changing stream hydraulics. Ground water reserves are becoming more and more depleted as surface water sources have become too polluted for human use.

It is important to appreciate the inextricably close relationship between land and water management. Water, which is a renewable resource, can in fact be put to good use only if the land on which it falls, and the land to which it is applied, are properly cared for. Land, which is for all purposes a non-renewable and inelastic resource, must be managed in such a manner as to be benefited rather than suffer damage as a result of its contact with water. The key to environmental quality, therefore, lies in scientific land and water management.

Biodiversity in the country's once extensive wetlands and coastal mangroves is fast declining. To add to this, current socio-economic activities and economic incentives are encouraging the unsustainable consumption of this resource.

The realization that India's water resources need to be more carefully managed, is leading to the adoption of sustainable water management practices. By managing its water more judiciously, India can avert the crisis that looms large over the future.

Water security is emerging as an increasingly important and vital issue for India. Many Indian cities are beginning to experience moderate to severe water shortages, brought on by the simultaneous effects of agricultural growth, industrialization and urbanization. These shortages would be further aggravated by receding of glaciers and dwindling fresh water resources.

Population stress, irrigation requirements and industrialization are the major pressures for water insecurity. The environmental challenges of water resource development and management in India are expected to manifest themselves more explicitly and rapidly in the coming years. These environmental challenges may be addressed through four broad approaches:

1. Improving efficiencies and minimizing losses.
2. Recharging ground water aquifers.
3. Abatement and treatment of water pollution.
4. Re-use and re-cycling of waste water.

8.2 Utilization of Water

The major use of water is for irrigation and hydropower generation, but water is also used for public water supply, industrial and navigation purposes. Water supply for irrigation can be obtained both from surface and ground resources.

Access to adequate water is one of the leading factors limiting development in India. Agricultural, industrial and domestic uses are competing more and more over a limited supply. Agricultural sector continues to dominate water use owing to its continued importance in the Indian economy. Industrial sector demand is increasing as the sector continues to grow at a faster rate. Domestic needs claim only a small portion of the annual water withdrawals as access to adequate water and sanitation supplies remains low throughout the country. Greater access and an improvement in the socio-economic situation are likely to result in a higher demand for water in the coming years, particularly in rural India.

8.2.1 Agricultural Demand for Water: Agriculture remains central to the Indian economy and, therefore, receives the greatest share of the annual water allocation. The necessity of irrigation in agricultural production is greater due to the unpredictable nature of the monsoon.

In regions completely dependent on rain-fed agriculture, a weak monsoon season can result in drought like conditions, leading to reduced yields or even total crop failure. The normal monsoon too allows the farmers to produce only one crop a year with a low yield. The productivity of irrigated agriculture per unit of land has been estimated at 7 times more than that of the rain-fed agriculture.

8.2.2 Industrial Demand for Water: In the past several decades, industrial production has increased in India owing to an increasingly open economy and greater emphasis on industrial development and international trade. Water consumption for this sector has consequently risen and will continue growing.

The major uses of water in industry are for cooling, processing and boiler feed. Industrial needs of water are increasing rapidly. Hence, the need to pay attention to methods of conservation and re-use of water in industries has become urgent. Most of the water used for industrial purposes is renewable in the sense that it becomes available for re-use if properly reconditioned.

Hydroelectric generation already accounts for a large percentage of water demand. The large untapped potential, particularly in the northern regions of India, and the growing demands for electricity from a larger population and industrial sector, will ensure that the development of this activity continues in the coming years.

8.2.3 Domestic Demand for Water: Demand from the domestic sector has remained low and accounts for only 5 percent of the annual fresh water withdrawals in India. Domestic water use will increase as the population continues to grow and access to water is improved. Most urban areas are serviced by a municipal water distribution system. Usually, the municipal water supply originates from local reservoirs or canals, but in some cases water may be imported through inter-basin transfer.

Although major cities in India enjoy access to central water supply systems, these schemes often do not adequately cover the entire urban population and are notoriously inefficient and unreliable. In rural areas, access to water is even more precarious. Over 80 percent of the rural domestic water comes from ground water sources since it is more reliable in terms of water quantity and quality. Still, in areas where water is scarce, rural women travel long distances to wells or streams to fetch water for their daily needs.

8.3 Ground Water Irrigation

The last three-four decades have witnessed a 'global boom' in groundwater use for irrigation in areas subject to extended dry seasons and/or regular droughts (Llamas, 2005). In India, for example, the groundwater-irrigated area has increased 500 percent since 1960 (Shah, 2009). According to a recent UN-FAO survey, irrigated agriculture is the largest abstractor and consumer of groundwater, with almost 40 percent of all cultivated land under irrigation being 'water well-equipped', with large groundwater-dependent agro economies in South and East Asia (Table 8.1). The nations with the largest groundwater-use areas are India (39 mha) and China (19 mha).

Agricultural land-use practices in general also exert a major influence on groundwater recharge quality (Foster and Candela, 2008) through the following:

1. **Leaching of Soil Nutrients:** This problem has been exceptionally widespread in the industrialised nations with (largely successful) attempts to increase grain, oilseed,

vegetable, fruit and milk production per unit area through the replacement of traditional crop rotations with near monocultures, but as yet has been less severe in the developing world where inorganic fertiliser applications have generally been much lower—in theory at least the problem of soil nutrient leaching should also be more manageable in irrigated than rain-fed agriculture.

Table 8.1: Global Survey of Groundwater Irrigation

Region	Groundwater Irrigation		Groundwater Volume Used	
	Mha	Total (percent)	km^3/a	Total (percent)
Global Total	112.9	38	545	43
South Asia	48.3	57.	262	57
East Asia	19.3	29	57	34
Southeast Asia	1	5	3	6
Middle East and North Africa	12.9	43	87	44
Latin America	2.5	18	8	19
Sub-Saharan Africa	0.4	6	2	7

Source: Siebert et al., 2010.

2. **Contamination with Pesticides:** A potentially serious problem but confined geographically to recharge areas of aquifers exhibiting high vulnerability to pollution from the land surface, where the more 'mobile' pesticides (mainly certain herbicides and soil insecticides) have been regularly used at high application rates.
3. **Mobilisation of Salinity:** This issue is of very serious concern in arid and hyper-arid areas where the 'irrigation frontier' has (or is) being extended through clearing of native desert scrubland with high salinity levels retained in the sub-soil profile.

In most regions that experience an extended dry season,

consumptive water use by agriculture (if unconstrained) usually generates a demand for crop irrigation in excess of the availability of renewable groundwater resources, given that extensive areas of cultivatable land usually occur above aquifers. Moreover, groundwater (a common-pool, open access resource) is also prone to the 'tragedy of the commons', with individual short-term interests prevailing over long-term communal concerns—and its effective management requires collective action (Ostrom, 1990; Burke and Moench, 2000). This situation has led to widespread depletion of groundwater resources, with the following collateral effects, which vary considerably in occurrence and intensity with hydrogeological setting:

- Counterproductive competition between irrigation users.
- Conflicts with rural and/or urban drinking water provision.
- Impacts on natural aquifer discharge (springflow, riverbed flows), which result cumulatively in an unacceptable impact on 'downstream' surface water flows.
- Degradation of important ground water dependent aquatic ecosystems.

8.4 Ground Water Depletion

Continuous groundwater depletion resulting from long-term excessive resource exploitation can in some cases result in a number of other serious consequences:

1. The salinisation of aquifers—which is a very insidious and often complex process arising from a variety of physical mechanisms.
2. Troublesome land subsidence due to the settlement of inter-bedded aquitards in alluvial and/or lacustrine formations increasing (and in some cases spiralling) electrical energy costs for pumping, especially where use is 'buffered' by subsidies or flat-rate tariffs—with serious implications for many electricity utilities and for the unit energy consumption and carbon footprint of irrigated agricultural production (Shah and Verma, 2008). India with diversified rainfalls and soil types faces a wide

variety of water related problems in agriculture. Runoff losses, ground water depletion, waterlogging, reservoir siltation and non-point source pollution are some of the key issues related to water in Indian agriculture. [1]

Increased irrigation for rice based cropping system further enhances the existing problem of *waterlogging* in coastal areas (Mishra, 1999). Coastal Odisha which mostly constitutes of alluvial plains when put to irrigation by the major river valley projects become highly *waterlogged* due to flow gravity—this is particularly noticed in the Paradeep to Chilka zone under the Mahanadi Delta Irrigation System Stage II. About 20 percent of the total area is waterlogged and about 1200 sq. km. has been moderately *waterlogged* due to agricultural activity. The total waterlogged area in the coastal area of the State is 60,50,000 ha and out of every 100 ha that have come under irrigation in the State, about 8 ha every year have gone out of cultivation due to over irrigation (Mishra, 1999).

A major portion of the estimated 9.38 million ha inflicted by *soil salinity* and *sodicity* in the country occurs in the irrigated land (GoI, 1990). The potentially saline soils occupy vast tracts in the black soil region and in the Indo-Gangetic plains and many of these areas were salinized when brought under canal irrigation. Prolonged use of poor quality water for irrigation can also cause salinity or sodicity. Table 8.2 shows the distribution of different talukas/blocks in India into overexploited and dark/critical with respect to the status of groundwater. The situation may not look serious at the national level since the number of such blocks is below 15 percent overall, and low in most states. However, the situation seems to be very precarious in states such as Punjab, Haryana, Karnataka and Tamil Nadu. In these locations, the incidence is high and the situation is critical.

Unit of assessment: Mandals—Andhra Pradesh; talukas—Maharashtra, Gujarat, Karnataka; Rest of the states—blocks. Overexploited: >100 percent; Dark/Critical: >85 percent but<100 percent.

Table 8.2: Extent of Ground Water Overexploitation

State	No. of Districts	No. of Blocks/Talukas/ Watersheds	Overexploited		Dark/Critical	
			No.	%	No.	%
Andhra Pradesh	22	1231	219	17.79	77	6.25
Bihar	42	589	6	1	14	2.4
Gujarat	20	223	31	13.9	12	5.19
Haryana	17	113	55	48.67	11	9.73
Karnataka	19	175	65	37.14	3	1.71
Kerala	14	151	5	3.31	15	9.93
Madhya Pradesh	45	312	24	7.69	5	1.6
Maharashtra	29	318	7	2.2	1	0.31
Punjab	17	137	103	75.18	5	3.65
Rajasthan	32	237	140	5.91	50	21.1
Tamil Nadu	27	385	142	36.88	33	8.57
Uttar Pradesh	74	820	39	4.76	13	1.59
West Bengal	16	341	0	0	61	17.9

Source: Central Ground Water Board (2006).

It has been estimated that in the poor quality areas of Haryana, the water table is rising by 0.15-1.0m per year (FAO, 1985). In Haryana, up to 0.4 million ha already has high ground water levels and up to 1.5 to 2 million hectares may be affected by high water levels and water logging and in the long run by soil salinization due to which a loss of ₹ 26.8 crore was estimated for the year 1982 and projected to rise to ₹ 71.9 crore by 2000 (Gangwar and Toorn, 1987; Joshi and Tyagi, 1991). In central-southern districts of Jind, Rohtak, Bhiwani, Hisar and Sirsa where most area is irrigated by canals, there is a very fast increase in water table (8-22 cm/year) leading to waterlogging (624 thousand hectares) and salinization (526 thousand hectare). Unfortunately, most of the underground water in these districts is sodic/saline, which makes the reclamation of these soils more difficult. Already 73 percent of the salt affected soils of the state fall in these districts (Bhargava, 1988).

The saline water belt of Punjab also shows similar trends, for e.g. in Muktsar block of Faridkot district, the water table rose from 1.7 meters to 2 meters (Khepar and Kaushal, 1991) (Joshi and Tyagi, 1991). Sehgal et al. (1985) noted that in Faridkot district of Punjab, the salt affected area increased from 6,000 ha in 1973 to about 82,000 ha in 1984. Gupta (1988) indicated that injury may develop on sensitive plants when irrigated with water containing boron in excess of 3 ppm in salt affected soils; boron may be to the extent of 10 ppm. Some of the calcareous alkali soils, particularly under waterlogged conditions contain excess of molybdenum. The plants may accumulate molybdenum in higher quantities, toxic to cattle, which results in diarrhoea, loss in body weight, and reduction in milk yield. This condition is known as "Teartness". Nayyar (1977) reported that 17 to 43 percent of berseem collected from different calcareous flood plains of Punjab had toxic levels of molybdenum.

8.5 Farm-level Effects of Irrigation-induced Salinity

The impact of salt-affected soils on crop yields is well documented, based on experimental trials. Table 8.3 shows the effect of irrigation water on crop yields under different cropping systems observed at the Chandra Sekhar Azad University of Agriculture and Technology Seed Farms in Kanpur, India. The data shows the sensitivity of different crop yielding ability to the intensity of irrigation and crop rotations. Pulses such as mung beans and pigeon peas were relatively more sensitive to salinity than cereal crops. In general, higher cropping intensity and irrigation frequency was associated with greater yield losses over the period 1974-75 to 1989-90.

There were a few comprehensive studies of farm-level effects of irrigation-induced salinity. In one study conducted by Joshi and Jha (1991) in the Sharda Sahayak irrigation project in India, the authors found that the yields of paddy and wheat were 41-46 percent lower on the degraded soils and net incomes in salt-affected lands were 82-97 percent lower than the unaffected land. Productivity losses were a result of increased costs of production: the per unit costs for paddy rose by about 60 percent, while for wheat, per unit costs increased by about 85 percent in saline lands. Using decomposition analysis, the study found that salinity accounted for as much as 72 percent of the difference in gross income between normal and salt-affected plots. The study also found that farmers reverted to low-input traditional varieties and practices as soil conditions deteriorated.

Table 8.3: Effect of Irrigation Water on Crop Yields

Crop Rotation (Annual Rainfall 765 mm)	Crops	No. of Irrigation/ Year	Crop Yield (t/ha)		Yield Reduc-tion (%)
			1974-75	1989-90	
Pigeon pea-fallow	Pigeon pea	1	0.15	0.12	20
Mung bean-Wheat	Mung bean	1	0.05	0.02	60
	Wheat	5	4	3	25
Rice-Wheat/Potato	Rice	6	4	3	25
	Wheat	5	4	2.5	38
	Potato	7	30	15	50

Source: Gupta and Abrol (2000), Table 6.

In salt affected areas, particularly where ground water is also saline, fluoride toxicity may be found. According to Tewari (1988) 20 percent of the area in India can be designated as endemic fluoride and about 5-8 percent of the Indian population is drinking water containing 1.5 mg/kg or more of fluoride. Gupta (1988) reported that the concentration of fluoride ranges from trace to >10 ppm in natural waters. The cases of fluorosis have been reported from Punjab, Haryana, Rajasthan, Bihar, M.P. and A.P. where soil, underground wares and plant materials contain excess of fluoride (Kanwar and Mehta, 1968; Susheela et al., 1987).

Due to the increased use of saline groundwater (containing high residual sodium carbonate) for irrigation purposes, soils develop a high pH of 8.9-9.8, the nutrient availability declines and soils acquire unusual hardness and surface crusting takes place in the arid regions. Seasonal cycles of puddling (wet tillage) and drying, over the long-term, leads to the formation of hardpans in paddy soils. The productivity of the wheat crop is affected by its poor establishment following puddled rice. It increases bulk density of surface soil with consequent decrease in infiltration rate and also water absorption in deeper soil layers, which restricts prolification of wheat roots. Large scale adoption of rice-based cropping systems has increased this problem in the Indo-Gangetic plain.

Some recent studies indicated either a decline or stagnation in yield of the intensive irrigated rice systems (Flinn and De Datta 1984; Cassman and Pingali, 1995, Nambiar, 1995, Pingali et al., 1997, Greenlands, 1997; Yadav, et al., 2000, Dawe et al., 2000; Kumar and Yadav, 2001). In most of these studies, the magnitude of yield decline was reported more for rice than for wheat—in fact a few studies reported increasing trend for wheat yields under irrigated ecosystem. These studies were largely based on experimental data designed with a specific objective under controlled environments (fixed nutrient doses, variety, other management practices, etc.) in the research farms and adaptive research trials.

Due to the increase in well irrigation, there has been a fall in the water tables in many areas. Bore irrigation in Punjab [2] and parts of Haryana has resulted in falling groundwater tables by more than a meter each year while canal irrigation in Uttar Pradesh, Bihar and West Bengal has led to waterlogging and salinisation. The groundwater resources in large parts of Punjab and Haryana of late have been showing signs of over development. Of the 118 development blocks in Punjab and

108 in Haryana, about 56 percent of the blocks in Punjab and 29 percent in Haryana have been classified as 'dark'. The water table in vast portions of the region has already reached a level of more than 15-20 meters below the ground surface and is declining at the rate of almost one meter per year. The over development of groundwater resources in this semi-arid region with an annual average rainfall of about 600 mm has in large part been the result of large scale cultivation of such water intensive crops as rice and wheat and installation of a large number of private tube wells. The area under rice and wheat have increased from 47 percent in 1970-71 to more than 70 percent in 1990-91 for Punjab and from 28 percent to 60 percent for Haryana during the same period. The large scale shifts in the cropping pattern towards rice-wheat have primarily been the result of its relative private profitability vis-à-vis alternative cropping systems (Malik, 1994).

In Punjab, there has been a significant decline in the area under waterlogging due to rapid growth in the utilization of groundwater through tube wells (Chopra, 1990; Ramesh Chand, 1996), but now the state is facing a serious problem of groundwater depletion. The water table is reported to be falling annually by 30-45 cm in the districts of Ludhiana, Patiala and Sangrur, which together contribute 30 percent of rice and wheat production in the state. The water table in Punjab is receding at a rate of 0.3 m to 0.5 m per year (Joshi and Tyagi, 1991).

There are evidences of increase in number of tube wells in Haryana where the increase has been 18-fold during the last 25 years resulting in 328 percent increase in the total area irrigated by tube wells in this region. Due to this over exploitation of groundwater in the North-eastern region of Haryana, there has been a lowering of water table at the rate of 12 to 33 cm per year (Chand and Haque, 1998). Such observations are made for Western U.P. as well where the water table was found falling by 50 to100 cm annually. Due to this, the farmers are compelled to lower their borings every third or fourth year (Ram and Sharma,

1997-98). In some upland areas of W.B. and Bihar, some farmers have started grossing rice-rice-wheat, which requires even more irrigation (Chand and Haque, 1998). Average fall in water tables was found to be one meter a year (Vaidyanathan, 1994). And in some places deep drilling has led to saltwater intrusion into aquifers, rendering water unfit for agricultural or domestic use (Kerr, 1996).

Availability and quality of the water resources are emerging problems in wheat cropping systems of the world. Quantity of water available for production is already a problem in most of South Asia due to the declining water tables (Harrington et al., 1993; Hobbs and Morris, 1995). Declining water tables not only raise production costs (by forcing farmers to pump water from greater depths), but also raise questions about the future availability of water resources to maintain current levels of productivity. Quality of irrigated water has also become a problem, especially in South Asia, since water-borne salts and minerals cause salinity and sodicity in soils. Salinity and sodicity causes hardening of the top soil, reduced plant stand (through poor seed emergence) and reduced water infiltration; all of which contribute to decrease in productivity. Singh et al. (1987) also indicated that nitrate contamination of groundwater in the Indian Punjab substantially increased water quality degradation in the past.

8.6 Water Use Efficiency

Among the factors explaining the low water use efficiency levels, the following are notable:

1. Non-provision of lining in canals.
2. Damage of lining in lined canals.
3. Poor maintenance of canal and distribution network, resulting in growth of weeds and vegetation within them.
4. Siltation of canals.
5. Distortion of canal sections due to siltation or collapse of slopes.
6. Leakages in gates and shutters.

7. Field channels and water courses passing through permeable soil strata, resulting in high seepage losses.
8. No regulation gates on head regulators, leading to uneven distribution of water.
9. Over-irrigation due to non-availability of control structures in the distribution system.
10. Poor management practices and lack of awareness among farmers.

Management of water resources poses increasingly difficult challenges. The total quantity of usable fresh water annually available in India is fixed, but its demand from expanding agriculture and other sectors is increasing. Water resources in many parts of the country are under severe stress leading to excessive exploitation of groundwater. There is some scope for increasing water availability. While these opportunities must be fully exploited, the real solution has to come from greater efficiency in water use.

Agriculture accounts for 80 percent of water needs at present, and there is considerable scope for increasing efficiency of water use in this area. This requires better management of water in areas of large and medium irrigation projects. It also requires putting in place more holistic aquifer management strategies. Separation of electricity feeders for agriculture and domestic use can help limit the availability of electricity for pumping groundwater thus breaking the vicious cycle between free energy and excess use of groundwater. Wherever this has been done in combination with large-scale watershed programmes, there is evidence that groundwater levels have recovered.

Normally, efficient use of scarce resources requires appropriate pricing, but pricing of water is a sensitive issue. This problem can be solved by providing *lifeline* water supplies for drinking and cooking at very low prices, while charging appropriately for additional water use by domestic consumers. There is a stronger case for rational pricing reflecting the scarcity of water for commercial and industrial use. There is also a strong case for rational pricing of water for agricultural purposes. The proportion of water recycled in urban areas, and by Indian industry needs to be significantly increased. This will happen if supply for commercial purposes is appropriately priced.

8.7 Strategy for Efficient Management of Water Resources

Emphasizing the need for a paradigm shift to meet the challenges of water scarcity, the Twelfth Five Year Plan (2012-17) observed, "These challenges can only be met through a paradigm shift in the management of water resources in India. This shift comprises the following elements:

- A move away from a narrowly engineering-construction-centric approach to a more multidisciplinary, participatory management approach in our major and medium irrigation projects, with central emphasis on command area development and a sustained effort at improving water use efficiency.
- Since groundwater accounts for nearly two-thirds of India's irrigation and 80 percent of domestic water needs, we need a participatory approach to sustainable management of groundwater based on a new programme of aquifer mapping.
- A massive programme for watershed restoration and groundwater recharge must be launched by transforming Mahatma Gandhi National Rural Employment Guarantee Act (MGNREGA) into our largest watershed programme, giving renewed energy to the reformed Integrated Watershed Management Programme launched in the Eleventh Five Year Plan and launching a completely revamped programme on repair, renovation and restoration (RRR) of water bodies.
- A new approach to rural drinking water and sanitation.
- All urban water supply projects to necessarily integrate sewage systems within them.
- Definite targets for recycling and reuse of water by Indian industry to move in conformity with international standards.
- Renewed focus on non-structural mechanisms for flood

management.

- Vastly improved systems of water-related data collection and management as also transparency in availability of data.
- Adaptation strategies to mitigate the likely impact of climate change to be pursued under the National Water Mission (NWM).
- Perennial rivers with sufficient draft through the year could be the focal point of a renewed thrust to inland waterways transport as an environment friendly economical mode of transport compared to road and rail.
- A new legal and institutional framework for water, based on broader consensus among the States". [3]

To sum up, water is a prime natural resource, a basic human need and a precious national asset. Planning, development and management of water resources need to be governed by national perspectives. Ecological concern is not merely an environmental concern but also calls for prudent use of water resources. Planning and implementation of water resources projects involve a number of socio-economic aspects and issues such as environmental sustainability, appropriate resettlement and rehabilitation of project-affected people and livestock, public health concerns of water impoundment, dam safety etc. Economic development has posed certain challenges for the planners and administrators, associated with the allocation and distribution of water resources.

There is large scope for improving water use efficiency in India. Agriculture consumes around 80 percent of available water resources at present and its water use efficiency is among the lowest in the world. Absence of rational pricing for canal water, combined with free or very cheap power for agriculture, has encouraged agricultural practices which are extremely wasteful. Cheap power has encouraged excess drawal of groundwater leading to falling water tables in large parts of the country. However, the man-made crisis of falling water tables is forcing some change as farmers are beginning to recognise the need to adopt technologies that economise on water.

Endnotes

1. Noted agriculture scientist M.S. Swaminathan submitted report on 'More Crop and Income per Drop of Water' in 2006. He has suggested formation of water users associations and training of two water masters each in 5,000 villages with the help of 50 water and farm institutes to create awareness about better management of ground water resources from this rabi season. Recommending that the year 2007-08 be declared as the "Year of more crop and income per drop of water", the farm expert said modern technology would have to be "married" with traditional wisdom to tackle the issue of depleting water resources, overuse of ground water and replenishing soil health. Groundwater level in Punjab region, for instance, was dipping at the rate of two to three feet per annum. The report is also available online at Ministry website, http://wrmin.nic.in/writereaddata/mainlinkFile/File722.pdf.
2. While doing a field survey in the villages of Punjab for the project for the FAO on Role of Agriculture (ROA) in 2003, I personally observed few interesting things. The production of water-guzzling paddy crop and spurt in the use of heavy-duty tubewells in the state led to drastic fall in water table from 8 cm to 95 cm per annum. This resulted into creation of a new capitalist system of private tubewell. There has been what Navroz Dubash (2002) describes as "a race to the bottom of the aquifer". Water table levels have plummeted to more than 100 feet (almost 30 metres) in the region. The traditional political economy has viewed agrarian institutions purely as a means of surplus extraction through the exercise of power, ignoring the role of the environment in shaping outcomes. The more fashionable new institutional economics approach assumes that institutional efficiency can overcome the problems of imperfect information availability. However, this provides a limited view. It fails to consider economic institutions as being embedded in social life, and which do not necessarily throw up the most efficient outcomes. Now the linkages of environmental degradation and political economy are much more evident in this situation. The overexploitation of ground waters in the state led to increase the cost of irrigation and so the cost of production. This creates a pressure on the political economy by demanding a hike in the minimum support prices (MSPs) fixed by the

Government of India.
3. Government of India, Planning Commission, *Twelfth Five Year Plan (2012-17),* Volume I, Chapter 5, para 5.5.

9

Summary and Conclusions

Agricultural production in India is mainly labour-intensive. In the literature, in a labour surplus economy like India, the average labour productivity determines agricultural wage rate (Ranis, 2004). In other words, in situations where land-man ratios are high and marginal productivity of workers is relatively low, wage determination is based on the principle of sharing output and/or income. This argument is invoked to explain wage formation in the agrarian labour market in South Asia, including India (Bardhan, 1984).

Other things being equal, higher output per worker should increase the demand for workers and result in an increase in labour compensation as long as the labour supply curve is not perfectly elastic. Some theoretical models suggest that the causality between the two may also work in the opposite direction. Akerlof (1982), for instance, argued that higher real wages lead to greater effort from workers. Additionally, higher real wages put upward pressure on labour costs and cause firms to substitute capital for labour, thereby increasing the marginal productivity of labour.

Further, we also observed the absence of a strong link between the real wage and labour productivity in some states. Bargaining power of workers may be an important factor in weakening the link between labour productivity and real wage. This needs to be investigated further at the micro level as well as for other crops and regions. The deviations of the real wage from the marginal productivity of labour have implications for the farms' profitability margins, and thus may also have repercussions for unemployment and employment creation. These hypotheses warrant further exploration. Additionally, the future research should focus on the following questions:

- To what extent the gap between real wage growth and agricultural labour productivity growth contribute to the farm and non-farm employment?
- Is there a long-term link between labour productivity and real wages, and does the recent period reflect a substantial deviation from equilibrium?
- Do the changes in labour market conditions affect real wage growth in emerging markets, and to what extent does the role of these factors differ in India?

There are strong linkages between power pricing, technology use, equity and efficiency in groundwater development. These have fostered intensive groundwater utilization. A sharp increase in groundwater use has been recorded, leading to over-exploitation of groundwater. The expansion of groundwater use has resulted in a speedy decline in the groundwater table in several parts of the country (Dhawan, 1995; Moench, 2003). The evidence indicates that the fall in the water table is quite rapid in water scarce regions. In Rajasthan, this decline is recorded at the rate of one to five metres per year in varying conditions (Desai and Namboodiri, 2009). The water productivity estimated by recent CACP Report, 2013-14, also presents an alarming picture.

Punjab consumes the highest quantity of water of 5,400 litre for producing one kg of rice in contrast to the lowest of 2,700 litre in West Bengal. It is evident that water productivity of Punjab at 19 kg/lakh litre is half of West Bengal's 37 kg/lakh litre. Equivalently, Punjab consumes almost double water compared to West Bengal for producing every kilogram of rice. Thus, Punjab is less efficient to the tune of 50 percent compared to West Bengal in terms of water productivity. It raises an issue of comparative advantage of where to propagate paddy cultivation.

Punjab is most efficient in terms of land productivity of paddy, but least efficient in terms of water productivity. In contrast, West Bengal is most efficient from water consumption for paddy cultivation. And if one considers the real cost of irrigation (massive power subsidy and falling water table in Punjab, which is two to three time more than in West Bengal), Punjab will become an even less efficient producer of rice vis-à-vis West Bengal (CACP, 2013-14; p. 54). [1]

Policy Implications

A potential concern for Indian policymakers has been the relatively poor farm performance of North Indian states, especially Punjab during the last decade of 2000s. As this region has deepened its specialization in grain production to meet policymakers' goal of wheat and rice self-sufficiency, it has witnessed a decline in TFP growth rate over the previous two decades. If this region is to continue providing the cereal grains consumed throughout India, greater emphasis should be placed on reversing the declining trend in cereal grain yield and TFP growth rates experienced over the previous decades. The empirical findings of the study also warrant some important policy imperatives on the technical augmenting perspective. There is a view that agriculture being a state subject, it is the states that have to bring about technical change (Ahluwalia, 2011).

However, historically, Green Revolution was successfully implemented during the 1970s period when a similar structure of government existed. What is required is that the Central government play a catalytic role when the Planning Commission consults each state government for the formulation of its annual and Five Year Plans. Also, the empirical result of this study suggests that the efficiency level has not been maintained. The decline in efficiency in fact has eroded part of the positive impact of the technical progress. For wheat production growth to sustain in the future, the Indian government might need to look more carefully into the factors that have caused such a serious decline in efficiency. In this regard, the increase of public investment in R&E to develop new varieties of seeds, better credit delivery system in agriculture (Acharya, 1997), extension of rural infrastructure facilities such as irrigation, roads, and etc., in lesser developed region are vital. Since India's agriculture is land scarce and labour surplus, its growth requires augmenting land and labour using new technology.

Desai et al. (2011) advocated a location-specific new technology based on integrated farming. Such technology will bring down the unit cost in real terms, which will most benefit the poor. Furthermore, the advocated strategy would require a big increase in public expenditure on agriculture research, education and extension; irrigation and flood control; watershed development; and rural infrastructure such as soil-testing laboratories, roads, electricity, and so on. It is also needed for inputs such as seeds, breeds, fertilisers, organic manure, cattle feed and modern small implements and equipment. The two categories are complementary and will improve the performance of each other.

Further, groundwater has rapidly emerged to occupy a dominant place in India's agriculture landscape and food security. It has become the main source of growth in irrigated area and now accounts for over 60 percent of the irrigated area in the country. It is estimated that now over 70 percent of India's food grain production comes from irrigated agriculture, in which groundwater plays a major role. With rapidly depleting groundwater levels and an erratic rainfall pattern, water is vastly becoming a scarce resource at the national level and therefore, the policies for the production pattern in the state should take into consideration the long-term sustainability of the water resources. For Punjab, we observed a negative association of groundwater exploitation with TFP growth with water guzzling crop, rice. From policy perspective, following policy recommendations can be made.

- Continuation of rice-wheat cropping system may exaggerate the problem of groundwater depletion, and therefore change in cropping pattern in favour of low water-consuming crops needs to be encouraged by providing price and non-price incentives.
- With the current institutional framework, an increase in

number of irrigation blocks considered overexploited at an alarming rate is not surprising. Under the existing property rights, groundwater is considered as an open resource in which farmers make private investment thinking that they have absolute rights to groundwater beneath their land, treating it as private property. This results in unchecked extraction of groundwater. In order to make the institutions more effective, there is an urgent need to define water rights. An integral approach at various levels is needed in general and through formal institutions in particular. Community-based action is also required for the efficient use of water resources in scarcity conditions by making effective, the working of informal institutions. The establishment of tradable private property right in groundwater would be a major institutional reform. This could also empower communities to establish rights over the water they manage and address the issues of efficiency, equity and sustainability (Desai and Namboodiri, 2009).

- Supply of free electricity to agriculture, particularly for irrigation purposes, has led to injudicious use of irrigation water. A proper price policy for electricity-use backed with proper information dissemination regarding optimum use of irrigation water needs to be introduced in the state. However, withdrawing subsidies can increase the cost of production, which will ultimately impact the political economy. Thus, it is imperative to reduce the cost of production.

Further, agriculture has a ubiquitous presence in India and has a number of wide-ranging and important environmental impacts, including soil erosion, loss of soil fertility, water pollution from nitrate and pesticide leaching, emissions of ammonia, carbon dioxide and other greenhouse gases, biodiversity loss, etc. These impacts impose costs upon society in terms of diminished welfare as well as in terms of direct damage and clean-up costs. We have tested the negative association of TFP growth of rice with power consumption, which had been used as a proxy for groundwater exploitation, in Punjab. The unavailability of the appropriate data at micro and macro level, limits the analysis to extend for the other states. The issue of soil degradation and groundwater pollution and depletion must be taken seriously by policymakers, as it presents the very real threat of limiting the future gains in agricultural productivity as evidenced in the study. There is a need for reliable data on various factors, which lead to soil and groundwater depletion.

Understanding the erosional process, assessment of areas at risk, prediction of future rates, causes of underground water contamination, extent of soil, land and water degradation, all depend on the current and past data sets. Data sets may be at various spatial scales, experimental plots, and these should cover longer periods. Thus, these costs could not be included in productivity measures despite their direct relationship with agricultural production. The agencies responsible for such data compilation like Ministry of Agriculture, Ministry of Environment and Forests, Ministry of Water Resources, Central Water Commission, Central Pollution Control Board, etc. should be properly equipped with personnel, who can collect, compile and scrutinize this data in time series to bring it to a usable form. There is a need for improving the quality of micro data on various aspects of farm production and marketing, which can advance our understanding of the drivers and hindrances of productivity growth. In the wake of globalization, agriculture will face major challenges as well as opportunities, which would need to be monitored and analysed for suitable policy adjustments.

It is useful to consider the political economy of policies that supported the innovation in agriculture. However, in the 1990s, these kinds of policies appear to have been carried to extremes with respect to incentives, while neglecting infrastructure and information. Thus, while power and water have been ever more subsidized, distorting cropping patterns and straining the states' environment, the infrastructure (storage, air transportation, marketing support, power) and information (seeds, irrigation technology) required for diversification into higher value-added crops have not kept pace. This book suggests a conceptual framework for guiding a restoration of policy balance, necessary for sustained growth as well as avoidance of environmental harm.

Endnote

1. Unchecked growth of groundwater exploitation has however led to unsustainable levels of groundwater use, while area under surface irrigation has remained stagnant despite the rapid growth in irrigation investments, phenomenon which the Planning Commission has acknowledged raises issues of investment strategy and quality (Shah, 1993, 2009).

Appendix

Appendix

Edited Extracts on Agriculture from Twelfth Five Year Plan (2012-17)

Although agriculture now accounts for only 14 percent of gross domestic product (GDP), it is still the main source of livelihood for the majority of the rural population. As such rapid growth of agriculture is critical for inclusiveness. Important structural changes are taking place within the sector and there are definite signs of improved performance. Agricultural growth has accelerated compared to the Tenth Plan and diversification is proceeding. The National Sample Survey Organisation (NSSO) data brings out that rural labourers are shifting to non-agricultural work, tightening the labour market in agriculture and putting pressure on farm wages.

However, dependence on agriculture remains unchanged among the rural self-employed whose average farm size continues to decline with population growth. This is also an ageing, more feminised population, whose educated young members are less likely to want to stay in farming. The viability of farm enterprise, mostly small farms, must therefore be a special area of Plan focus in the Twelfth Plan. The Plan must also focus on other priorities such as resource-use efficiency and technology to ensure sustainability of natural resources, adaptation to climate change and improvements in total factor productivity.

Agriculture and Allied Sectors

The average of annual growth rates of GDP in agriculture and allied sectors during the Eleventh Plan is now placed at 3.3 percent. This is short of the target of 4 percent but is significantly better than the achievement of 2.4 percent in the

Tenth Plan. Failure to reach the target growth is one reason for the high inflation in prices of food and other primary commodities that persist despite the recent slowdown in overall GDP growth. Consequently, although the overall GDP growth target of the Twelfth Plan has been revised down since the Approach Paper, the growth target for agriculture is maintained at 4 percent.

The climate challenge facing agriculture needs to be taken seriously. A distinct trend towards both drier and warmer weather, in context of agriculture has traditionally been discussed in terms of the monsoon (that is, June-September) but annual precipitation is probably much more relevant now since the dominance of Kharif crops has reduced. Viewed in this perspective, it is noteworthy that each of the last three Plan periods has recorded lower mean rainfall and higher rainfall variability compared to the immediately previous period. Three (2008, 2009 and 2011) of the five Eleventh Plan years had annual rainfall below 95 percent of long period average, as compared to only five in the previous 15 years. Temperature conditions have deteriorated even more. Periods prior to 1997 can be considered normal, but warming has increased at an accelerating pace since then. The Eleventh Plan period contained the two warmest years (2010 and 2009) ever recorded since 1900. Even the coolest year (2008) during these five years was the thirteenth warmest in the last 110 years.

Livestock contributes 25 percent of gross value added in the agriculture sector and provides self-employment to about 21 million people. Rapid growth of this sector can be even more egalitarian and inclusive than growth of the crop sector because those engaged in it are mainly small holders and the landless. Growth of livestock output averaged 4.8 percent per annum during the Eleventh Plan recovering from an average of 3.6 percent in the Ninth and the Tenth Plans.

Growth of dairying, which is the main constituent of livestock sector, though slightly higher than the 4 percent averaged since 1990, was short of demand. With over 75 percent of cattle located in rain-fed areas, the major issue is access to feed, fodder and drinking water which is becoming increasingly scarce. The problems of the sector are compounded by growing numbers of unproductive male cattle. Developing a strong fodder base needs intensive effort and innovation in institutional aspects of pasture protection and management and usufruct sharing. There is little concerted effort in this area at present as it is too fragmented across various departments to be able to provide the technical inputs, institutional designs and adequate investments to make a meaningful impact. Richer farmers with access to groundwater irrigation can grow irrigated fodder and increase herd size. Poorer livestock owners, dependent mainly on commons and agriculture residues, end up underfeeding the animals. This problem raises questions about the present breeding strategy that focuses almost exclusively on induction of breeds that are high yielding, but are much less tolerant to adverse conditions in extensive livestock systems.

These issues, which also affect owners of small ruminants, poultry and even those involved in inland fishery, came to the fore during the Eleventh Plan following the drought of 2009. The consequent high inflation in feed and fodder, that also led to high inflation in prices of livestock products, revealed a need for much greater coordination not only between agencies responsible for livestock and those responsible for crops that sustain livestock, but also with other policies, for example, trade policies that influence feed and livestock product prices. RKVY provided a window which cut across departments to allow States to focus on fodder shortages and restored growth of livestock output much quicker than in earlier droughts. Nonetheless, underlying problems remain, as does so called protein inflation. The Twelfth Plan must address these problems by involving dairy cooperatives in breed and feed issues, revisit breeding strategies and make fodder development higher priority in both animal husbandry and crop programmes.

India produces about 65 billion eggs annually and

production growth has accelerated from around 4 percent per annum during the 1990s to over 5 percent during the Tenth and the Eleventh Plan. This acceleration has been achieved despite new challenges such as periodic outbreaks of avian influenza and the bio-fuels effect on international prices of maize, the main poultry feed, which has now transmitted into the domestic economy. One reason for this vitality has been the growth of a large and vibrant commercial poultry sector with adequate economies of scale and fairly good backward and forward linkages. Besides eggs, this commercial poultry sector also produces over 2 million tonnes of broiler meat which is an increasing part of total meat production of about 5 million tonnes. Meat, with production growth at over 5.5 percent per annum during the Eleventh Plan, is the fastest growing segment in the livestock sector.

The performance of the fisheries sub-sector has been impressive on the whole, with growth more than 5 percent per annum during the 1980s and 1990s, but growth in this sub-sector has been decelerating since mid-1990s. The main reason for this has been stagnation of marine fishery, a phenomenon which is expected to continue. The major growth in fisheries in recent years has come from the inland fisheries, with particularly rapid development of brackish water aquaculture. This has been linked to prawn cultivation for export, although there is also strongly growing domestic demand for fresh water fish. Fish prices more than doubled during the Eleventh Plan, a higher inflation than either crops or any other livestock segment, despite a small acceleration in production growth compared to the Tenth Plan. A problem in this sector is that although a National Fisheries Development Board was set up, responsibilities are still not clearly defined between this and the Department of Animal Husbandry, Dairying and Fisheries. This has in particular meant an inability to realise the vast potential of inland fresh water fishery. Fish production can be enhanced 2 to 4 times in rain-fed water bodies, whether through irrigation reservoirs, natural wetlands or ponds and tanks created by watershed development under Mahatma Gandhi National Rural Employment Guarantee Scheme (MGNREGS). If fully harnessed, these can secure over 6 percent fishery growth in the Twelfth Plan.

A very important reason why subsidies should be rationalised and restrained is that some of these subsidies could actually be doing harm. A case for subsidies exists if there is clear evidence that some input is being underused. Conversely, when there is clear evidence of overuse of a subsidized input, there is a case to reduce or even eliminate the subsidy. Today, there is clear evidence of overuse. Data from all over India, especially from the prime Green Revolution areas, show that high use of chemical fertilisers and power is causing excessive mining of other soil nutrients and of groundwater, and that this is also leading to loss of quality of both soil and water. There is of course about 20-25 percent of the country's arable area, located largely in North-East, East and Central India, where use of these inputs is so low that further intensification is desirable *per se*. But with nearly 90 percent of fertilisers and 95 percent of farm electricity currently being used outside this area, there can be no doubt that the present subsidies are actually encouraging practices that need to be discouraged.

National Project on Management of Soil Health and Fertility (NPMSH&F)

Any proposal for reducing subsidies will be opposed by farmers on the grounds that output will fall if the subsidy cut reduces input use. This is true unless other investments are made simultaneously but such investments would indeed be facilitated by the resources released. Efforts were made in the Eleventh Plan to encourage more efficient practices without actually reducing the quantum of subsidy. For example, many States have undertaken separation of feeders so that electricity supply for agricultural use can be treated differently from that for rural non-agricultural use, and stricter scheduling imposed on the former while maintaining its lower price. Similarly, the Centre introduced a new scheme, the NPMSH&F to promote soil testing and issue of soil health cards to farmers, aimed particularly to spread awareness of micronutrient deficiencies resulting from excessive and unbalanced fertiliser use and to encourage balanced and judicious use of chemical fertilisers in conjunction with organic manures to maintain soil health and fertility. Moreover, in order to rationalise fertiliser subsidies, a nutrient-based subsidy (NBS) system was adopted to subsidise fertiliser products uniformly on basis of nutrient content, rather than set product-wise subsidies and separate maximum retail prices (MRPs) for each product. The objective was to reduce deadweight of the fertiliser control order, set nutrient-specific subsidies that maintain desirable NPK balance, and evolve a subsidy protocol to encourage both development of new complex fertiliser products (including micronutrients) and more investment in the sector.

These initiatives have had some success in particular regions, but they do not as yet show up in national data in terms of higher additional output per unit additional use of these inputs. Moreover, NBS roll-out was seriously flawed since urea was kept out of its ambit. Urea prices remain controlled with only a 10 percent rise at the time of adoption of the NBS in 2010. Meanwhile, prices of decontrolled products doubled. The fixity of the urea price naturally worsened the NPK balance. Also, there has been very little product innovation. The subsidy bill has increased because resulting higher urea demand has been met entirely by imports at a unit subsidy twice that on domestic output, with little incentive to expand domestic capacity. The NBS as rolled out has been counterproductive because urea has not been included.

Major Challenges and Priorities during the Twelfth Plan

The main lesson from the performance in the Eleventh Plan is that while there has been a welcome turnaround from the deceleration that was evident in the decade to 2005, and while several indicators have shown marked improvement and potential to build upon, several policy imbalances exist that can prove to be major handicaps. There are also other formidable challenges, for example, a shrinking land base, dwindling water resources, the adverse impact of climate change, shortage of farm labour, and increasing costs and uncertainties associated with volatility in international markets. The Twelfth Plan will need to face these challenges boldly.

The key drivers of growth will remain as under:

1. Viability of farm enterprise and returns to investment that depend on scale, market access, prices and risk.
2. Availability and dissemination of appropriate technologies that depend on quality of research and extent of skill development.
3. Plan expenditure on agriculture and in infrastructure which together with policy must aim to improve functioning of markets and more efficient use of natural resources.
4. Governance in terms of institutions that make possible better delivery of services like credit, animal health and of quality inputs like seeds, fertilisers, pesticides and farm machinery.

In addition, certain regional imbalances must be clearly addressed. A national priority from view of both food security and sustainability is to fully extend Green Revolution to areas of low productivity in the eastern region where there is ample ground water, and thereby help reduce water stress elsewhere. Rain-fed areas continue to be at a disadvantage, and their development still requires some mindset changes.

Small and marginal farmers face problems not only with shrinking land assets and with credit; they have difficulty in accessing critical inputs for agriculture such as quality seeds and timely technical assistance. In this situation, FPOs offer a form of aggregation that leaves land titles with individual producers and uses the strength of collective planning for production, procurement and marketing to add value to members' produce through pooled resources of land and labour, shared storage space, transportation and marketing facilities. These also improve bargaining power of small farmers and, most importantly, reduce transactions costs of banks and buyers to deal them. Investing in such group efforts has strong externalities.

The extension system of State agricultural departments is the weakest link in the chain between research and the farmer. Large number of vacancies of extension workers in the State Agriculture Department was one of the gravest concerns expressed by the Eleventh Plan document. During the Eleventh Plan, efforts were initiated to improve extension services by extending Central support to State extension reforms. This has resulted in 604 Agriculture Technology Management Agencies (ATMAs) to be established across the country with 21,000 new posts sanctioned with Central assistance at State, district and block levels. Also, since a continuous problem plaguing extension has been lack of organic link between the research system and the extension machinery, R&D linkage guidelines were jointly brought out by the DAC and ICAR and sent to all States and SAUs. The basic thrust of these guidelines were to get ATMAs and KVKs to work together at the district level and below, keeping in view the priorities reflected in Comprehensive District Plans. Although neither has delivered full results, there is now much greater acceptance that things must be done together.

Mahatma Gandhi National Rural Employment Guarantee Act (MGNREGA)

The Eleventh Plan period saw major new initiatives towards inclusive growth in rural India. The total budgetary allocation for all rural development programmes by the Government of India in the Eleventh Plan was ₹ 2,91,682 crore which accounted for 25 percent of the total Central Budget Plan provision. Rural development programmes cover employment through the Mahatma Gandhi National Rural Employment Guarantee Act (MGNREGA) and the National

Rural Livelihoods Mission; housing via the Indira Awaas Yojana and other State schemes and bank support; sanitation through the Total Sanitation Campaign; provision of drinking water via the National Rural Drinking Water Programme; social security through the National Social Assistance Programme; watershed development via the Integrated Watershed Management Programme; road connectivity through the Pradhan Mantri Gram Sadak Yojana; and electrification via the Rajiv Gandhi Grameen Vidyutikaran Yojana.

Over the last six years, MGNREGA has delivered the largest employment programme in human history, which is unlike any other in its scale, architecture and thrust. Its bottom-up, people-centred, demand-driven, self-selecting, rights-based design is new and unprecedented. Never have in such a short period so many crores of poor people benefited from a Government programme.

In 2011-12, nearly 5 crore families were provided over 211 crore person-days of work under the programme. Over the last six years, MGNREGA has generated more than 1,200 crore person-days of work at a total expenditure of over ₹ 1,66,760 crore. The share of SC/ST families in the work provided under MGNREGA has been 55 percent and 45 percent of workers are women. Average wages of workers have gone up by 54 percent over the last five years and wages have now been so indexed that workers will be protected from the ravages of inflation. Nearly 10 crore bank/post office accounts of our poorest people have been opened and around 80 percent of MGNREGA payments are made through this route, an unprecedented step in the direction of financial inclusion.

In many parts of the country, spectacular successes have been recorded in water harvesting. Distress migration has been arrested in several areas. Some State Governments have been leaders in this and the National Consortium of Civil Society Organisations on MGNREGA has also set up examples of excellent work.

Many critics and sceptics of MGNREGA who were extremely vocal during the years leading up to its passage by Parliament and in the early years of its implementation have been silenced, especially after it was recognised that the purchasing power the programme created in rural areas and the operation of the Keynesian multiplier played a crucial role in generating demand for industry during the dark days of the recession and assisted in our comparatively faster emergence out of it.

However, there is no denying the fact that the true potential of MGNREGA as an instrument of rural transformation is yet to be fully realised. Since the programme marks a radical departure from earlier efforts of a similar kind, there have been many problems in infusing the system with the new culture of demand-driven, rights-based, decentralized decision-making. The MGNREGA provides a historic opportunity for strengthening Panchayati Raj in India but the experience so far also alerts us to the need for doing much more in this direction.

There are problems that arise from the sheer scale of the programme. At the same time, new opportunities have arisen because of advances in information technology that allow us to get rid of inefficiencies and corruption in a manner quite inconceivable in the past. The MIS currently used by MGNREGA is already the best we have ever had. More than 8 crore muster rolls and over 12 crore job cards have been placed online. But there is huge scope for further improvement in overcoming systemic delays as shown by the software being used, for example, in Andhra Pradesh.

Bibliography

Bibliography

Abel, Martin E. and K. W. Easter (1971): "Agricultural Development Planning and Program Evaluation with a Focus on Regional Restraints", *Economic and Political Weekly*, Vol. VI, Nos. 30-32, July, Special issue.

Abler, David G., S. Tolley and G.K. Kripalani (1994): "Technical Change and Income Distribution in Indian Agriculture", Westview Press.

Acharya, S.S. (1997): "Agricultural Price Policy and Development: Some Facts and Emerging Issues", *Indian Journal of Agricultural Economics*, 52 (1); pp. 1-47.

Ackermann, Richard (2012): "New Directions for Water Management in Indian Agriculture", *Global Journal of Emerging Market Economies*; 4(2); pp. 227-288; Emerging Markets Forum; Sage Publications.

Adkins, Lee C., Ronald J. Moonmaw and Andreas Savvides (2002): "Institutions, Freedom, and Technical Efficiency", *Southern Economic Journal*, 69; pp. 92-108.

Afriat, S. (1972): "Efficiency Estimate of Production Functions", *International Economic Review*, 13; pp. 568-598.

Aghion P. and Howitt, P. (1998): "Endogenous Growth Theory", The MIT Press, Cambridge, Massachusetts.

Ahluwalia, Montek S. (2011): "Prospects and Policy Challenges for the Twelfth Plan", *Economic and Political Weekly*, XLVI (21), May 31; pp. 88-105.

Ahmad, Munir, Ghulam Mustafa Chaudhry and Mohammad Iqbal (2002): "Wheat Productivity, Efficiency, and Sustainability: A Stochastic Production Frontier Analysis", *The Pakistan Development Review*, 41:4 Part II (Winter 2002); pp. 643-663.

Aigner, D.J., C.A.K. Lovell and P. Schmidt (1977): "Formation and Estimation of Stochastic Frontier Production Function Models", *Journal of Econometrics*, 6; pp. 21-37.

Ali, M. and D. Byerlee (1991): "Economic Efficiency of Small Farmers in a Changing World: A Survey of Recent Evidence", *Journal of International Development*, 3(1); pp. 1-27.

Ali, M. and M.A. Chaudhry (1990): "Inter-regional Farm Efficiency in Pakistan's Punjab: A Frontier Production Study", *Journal of Agricultural Economics*, 41; pp. 62-74.

Arnade, Carlos (1998): "Using a Programming Approach to Measure International Agricultural Efficiency and Productivity", *Journal of Agricultural Economics*, 49 (1); pp. 67-84.

Bakhshoodeh, M. and K. Thomson (2001): "Input and Output Technical Efficiencies of Wheat Production in Kerman, Iran", *Agricultural Economics*, 24; pp. 307-313.

Ball, V.E., J.C. Bureau, R. Nehring and A. Somwaru (1997): "Agricultural Productivity Revisited", *American Journal of Agricultural Economics*, 79(4); pp. 1045-1063.

Ball, V.E., R. Färe, S. Grosskopf, O. Zaim and R. Nehring (2001): "Accounting for Bads in the Measurement of Productivity Growth: A Cost Indirect Productivity Measure and its Application to US Agriculture", AAEA and CAES Annual Meeting, August 5-8.

Banerjee, Abhijit Vinayak, Paul J. Gertler and Maitreesh Ghatak (2002): "Empowerment and Efficiency: Tenancy Reform in West Bengal", *Journal of Political Economy*, 110(2); pp. 239-280.

Banerjee, Abhijit Vinayak (2008): "Big Answers for Big Questions: The Presumption of Growth Policy", Brookings conference on What Works in Development? Thinking Big and Thinking Small.

Barah, B.C. (2005): "Dynamic of Rice Economy in India: Emerging Scenario and Policy Options", Occasional Paper 47; Department of Economic Analysis and Research, NABARD, Mumbai, India.

Barbier, E.B. (1989): "Cash Crops, Food Crops, and Sustainability: The Case of Indonesia", *World Development*, 17(6); pp. 879-895.

Bardhan, Pranab (1999): "The Political Economy of Development in India", Oxford University Press, New Delhi.

Bardhan, Pranab (2001): "Sharing the Spoils: Group Equity, Development and Democracy", in Atul Kohli (ed.), The Success of India's Democracy, Cambridge University Press: Cambridge, U.K.

Bardhan, Pranab (2003): "The Politics of Economic Reform in India", in *Poverty: Agrarian Structure and Political Economy In India*, Oxford University Press: New Delhi.

Bardhan, Pranab (2005): "The Nature of Opposition to Economic Reforms in India", *Economic and Political Weekly*, XL(48); November 26; pp. 4995-4998.

Barnum, H.N. and L. Squire (1978): "Technology and Relative Economic Efficiency", Oxford Economic Papers, 30; pp. 184-198.

Barro, Robert C. and Xavier Sala-i-Martin (1995): "Economic Growth", McGraw-Hill: New York.

Barua, Alokesh and S. K. Das (1996): "Regional Disparity, Economic

Growth and Liberalization: A Study of the Indian Economy", *Journal of Development Studies*, 32(3); pp. 364-390.

Basant, Rakesh (1987): "Agricultural Technology and Employment in India: A Survey of Recent Research", *Economic and Political Weekly*, 22; pp. 1297-1308.

Battese, G.E. and T.J. Coelli (1992): "Frontier Production Function, Technical Efficiency and Panel Data: With Application to Paddy Farmers in India", *Journal of Productivity Analysis*, 3; pp. 153-169.

Battese, G.E. (1992): "Frontier Production Function and Technical Efficiency: A Survey of Empirical Application in Agricultural Economics", *Agricultural Economics*, 7; pp. 185-208.

Bauer, P.W. (1990): "Recent Developments in the Econometric Estimates of Frontiers", *Journal of Econometrics*, 46; pp. 39-56.

Becker, Charles, M., Jeffrey G. Williamson and Edwin S. Mills (1992): "Indian Urbanisation and Economic Growth since 1960s", Johns Hopkins University Press, Baltimore.

Berndt, E. R. and L. R. Christensen (1973): "The Translog Function and the Substitution of Equipment, Structures and Labour in US Manufacturing, 1929-68", *Journal of Econometrics*, 1(1); pp. 81-114.

Bernstein, H. (2010): "Class Dynamics of Agrarian Change", Fenwood Publishing.

Bhaduri, Amit (1973): "A Study in Agricultural Backwardness under Semi-Feudalism", *Economic Journal*, 83(329); pp. 120-137.

Bhalla, G.S. (2006): "Condition of Indian Peasantry", National Book Trust, India, New Delhi.

Bhalla, G.S. and Gurmail Singh (2001): "Indian Agriculture: Four Decades of Development", Sage Publications, New Delhi.

Bhandari, A.L., R. Amin, C.R. Yadav, E.M. Bhattarai, S. Das, H.P. Aggarwal, R.K. Gupta and P.R. Hobbs (2003): "How Extensive are Yield Declines in Long-Term Rice-Wheat Experiments In Asia?", *Field Crops Research*, 81; pp. 159-180.

Bharadwaj, Krishna (1974): "Production Conditions in Indian Agriculture", Cambridge University Press, Cambridge.

Bhatia, B.M. (1988): "Indian Agriculture: A Policy Perspective", Sage Publications, New Delhi.

Bhatia, S.S. (1967): "A New Measure of Agricultural Efficiency in Uttar Pradesh, India", *Economic Geography*, 43(3); pp. 244-260.

Bhushan, Surya (2005): "Total Factor Productivity Growth of Wheat in India: A Malmquist Approach", *Indian Journal of Agricultural*

Economics, January-March, 60(1); pp. 32-48: Mumbai, India.

Bhushan, Surya (2006): "Productivity and Efficiency of Wheat Production in India", *Artha Vijnana*, XLVIII(4); pp. 305-320.

Bhushan, Surya (2009): "Non-Parametric Productivity Growth of Wheat in India during Post-Green Revolution", *The Indian Economic Journal*, January-March; 56(4); pp. 55-77.

Binswanger-Mkhize and P. Hans (2013): "The Stunted Structural Transformation of the Indian Economy: Agriculture, Manufacturing and the Rural Non-Farm Sector", *Economic and Political Weekly*, Review of Rural Affairs; Supplement June 29, XLVIII; 26 and 27: pp. 5-13.

Birthal, P.S. (2003): "Economic Potential of Biological Substitutes for Agrochemicals," Policy Paper 18, National Centre for Agricultural Economics and Policy Research, New Delhi.

Birthal, P.S. and D. Jha (1997): "Socio-Economic Impact Analysis of Integrated Pest Management Programmes", in National Symposium on IPM in India: Constraints and Opportunities, Indian Agricultural Research Institute, New Delhi. October 23-24.

Bliss, G.J. and N.H. Stern (1982): "Palanpur: The Economy of an Indian Village", Oxford: Clarendon Press.

Blyn, George (1966): "Agricultural Trends in India; 1891-1947: Output, Availability and Productivity", Pennsylvania University Press, Philadelphia.

Brar, K.K. (1999): "Green Revolution: Ecological Implications", Dominant Publishers and Distributors, Delhi, India.

Bravo-Ureta, B. E. and A.E. Pinheiro (1993): "Efficiency Analysis of Developing Country Agriculture: A Review of the Frontier Function Literature", *Agricultural and Resource Economics Review*, 22; pp. 88-101.

Brooks, Karen, J. Luis Guasch, Avishay Braverman and Csaba Csaki (1991): "Agriculture and the Transition to the Market", *Journal of Economic Perspectives*, 5(4); pp. 149-161.

Brown, L.R. (1997): "The Agricultural Link: How Environmental Deterioration Could Disrupt Economic Progress?", World Watch Paper No 136, World Watch Institute, Washington D.C., USA.

Burke, J. and M. Moench (2000): "Groundwater and Society: Resources, Tensions and Opportunities–Themes in Groundwater Management for the 21st Century", UN Department of Economic and Social Affairs, New York, USA.

Burmeister, E. (2000): "The Capital Theory Controversy", in Kurz, H.D (ed.), *Critical Essays on Pierro Sraffa's Legacy in Economics*, Cambridge University Press, Cambridge.

Byerlee, D. (1992): "Technical Change, Productivity, and Sustainability in Irrigated Cropping Systems of South Asia: Emerging Issues in the Post-Green Revolution Era", *Journal of International Development*, 4(5); pp. 477-496.

Byerlee, D. (1996): "Modern Varieties, Productivity and Sustainability: Recent Experience and Emerging Challenges", *World Development*, 24(40); pp. 697-718.

Byerlee, D. and R. Murgai (2001): "Sense and Sustainability Revisited: the Limits of Total Factor Productivity Measures of Sustainable Agricultural Systems", *Agricultural Economics*, 26; pp. 227-236.

CACP (March 2013): "Price Policy for Kharif Crops: The Marketing Season 2013-14", Department of Agriculture and Cooperation, Ministry of Agriculture, Government of India, New Delhi.

Capalbo, Susan M. and John M. Antle (eds.) (1988): "Agricultural Productivity: Measurement and Explanations", Resources for the Future, Washington, D.C.

Carson, R. (1962): "Silent Spring", Boston, Houghton Mifflin., reprinted in Nico Nellissen, Jan van der Straaten and Leon Klinkers (eds.) (2001), *Classics in Environmental Studies: An Overview of Classic Texts in Environmental Studies*; Kusum Publishing, New Delhi; pp. 68-73.

Cassman, K.G. and P.L. Pingali (1995): "Extrapolating Trends from Long-term Experiments to Farmers' Fields: The Case of Irrigated Rice System in Asia", in Barnett, Payne and Steiner (eds.), *Agricultural Sustainability: Economic, Environmental and Statistical Considerations,* John Wiley & Sons Ltd., U.K.; pp. 63-84.

Caves, Douglas W., Laurtis R. Christensen and W. Erwin Diewart (1982): "The Economic Theory of Index Numbers and the Measurement of Input, Output, and Productivity", *Econometrica,* 50(6); pp. 1393-1414.

Chakravarty, Sukhamoy (1984): "Power Structure and Agricultural Productivity", in Meghnad Desai, Susanne Hoeber Rudolph, Ashok Rudra (eds.), *Agrarian Power and Agricultural Productivity in South Asia,* Oxford University Press, Delhi.

Chambers, Robert (1984): "Beyond the Green Revolution: A Selective Essay", in *Understanding Green Revolutions Agrarian Change and Development Planning in South Asia,* Tim Bayliss-Smith and

Sudhir Wanmali (eds.), Cambridge University Press.

Chambers, Robert, G. (1988): "Applied Production Analysis: A Dual Approach", Cambridge University Press, Cambridge.

Chand, Ramesh (1996): "Agricultural Diversification in Punjab: Potentials and Prospects", Institute of Economic Growth, Delhi.

Chand, Ramesh (2004): "Impact of Trade Liberalization and Related Reforms on India's Agricultural Sector, Rural Food Security, Income and Poverty", Institute of Economic Growth, New Delhi.

Chand, Ramesh and Haque T, (1998): "Rice Wheat Crop System in Indo-Gangetic Region: Issues Concerning Sustainability", *Economic and Political Weekly*, 33(26); June 27; pp. A108-A112.

Chand, Ramesh, Praduman Kumar and Sant Kumar (2011): "Total Factor Productivity and Contribution of Research Investment to Agricultural Growth in India", Policy Paper 25, NCAP, New Delhi.

Chandrasekhar, C.P. and Jayati Ghosh (2006): "The Market That Failed a Decade of Neo-liberal Economic Reforms in India", Left World, New Delhi.

Charnes, A., W.W. Cooper and E. Rhodes (1978): "Measuring the Efficiency of Decision Making Units", *European Journal of Operational Research*, Vol. 2; pp. 429-444.

Charnes, A., W.W. Cooper, A. Lewin and L. Seiford (1994): "Data Envelopment Analysis: Theory, Methodology and Applications", Kluwer Academic Publisher.

Chennareddy, V. (1967): "Production Efficiency in South Indian Agriculture", *Journal of Farm Economics*, 49; pp. 816-820.

Chopra, K. (1989): "Land Degradation: Dimensions and Causalties", *Indian Journal of Agricultural Economics*, 44; pp. 233-241.

Chopra, K. (1990): "Agricultural Development in Punjab: Issues in Resource Use and Sustainability", Vikas Publishing House, New Delhi.

Cleaver, H.M. (1972): "Contradiction of the Green Revolution", *American Economic Review*, 62 (1/2); pp. 177-186.

Cleveland, W.S. and C.L. Loader (1996): "Smoothing by Local Regression: Principles and Methods", in W. Haerdle and M.G. Schimek (eds.), Statistical Theory and Computational Aspects of Smoothing, pp. 10-49, Springer, New York, accessed on 10th October, 2013.

Coelli, T.J. (1995): "Recent Developments in Frontier Modelling and Efficiency Measurement", *Australian Journal of Agricultural Economics*, Vol. 39 (3); pp. 219-245.

Coelli, T. J. (1996): "A Computer Program for the Stochastic Frontier Production and Cost Function Estimation", Working Paper 96/07, Centre for Efficiency and Productivity Analysis, University of New England, Australia.

Coelli, T.J. (1996): "A Guide to DEAP Version 2.1: A Data Envelopment Analysis (Computer Program)", Working Paper 96/08, Centre for Efficiency and Productivity Analysis, University of New England, Australia.

Coelli, T.J. and D.S. Prasada Rao (2003): "Total Factor Productivity Growth in Agriculture: A Malmquist Index Analysis of 93 Countries, 1980-2000", Working Paper Series, No. 02/2003, Centre for Efficiency and Productivity Analysis, School of Economics University of Queensland, Australia.

Coelli, T.J., D.S. Prasada Rao and George T. Battese (1998): "An Introduction to Efficiency and Productivity Analysis", Kluwer Academic Publishers, Boston.

Coelli, T.J., D.S. Prasada Rao, C.J. O'Donnell and George T. Battese (2005): "An Introduction to Efficiency and Productivity Analysis", Second Edition, Springer Publishers, USA.

Coelli, T.J. and G.E. Battese (1996): "Identification of Factors which Influence the Technical Efficiency of Indian Framers", *Australian Journal of Agricultural Economics*, 40(2), pp. 19-44.

Cohen, A. and G. Harcourt (2003): "Whatever Happened to the Cambridge Capital Controversies?", *Journal of Economic Perspectives*, 17, pp. 199-214.

Cones, Richard (1992): "Duality and Modern Economics", Cambridge University Press, London.

Conway, G.R. and J.N. Pretty (1991): "Unwelcome Harvest: Agriculture and Pollution", Earthscan Publications, London.

Conway, R.G. and B.E. Barbier (1990): "After the Green Revolution: Sustainable Agriculture for Development", Earthscan Publications, London.

Copeland, Morris A. (1937): "Concepts of National Income", Vol. 1, pp.3-63, *National Bureau of Economic Research*, New York.

Copeland, Morris A., and E.M. Martin (1938): "The Correction of Wealth and Income Estimates for Price Changes", Vol. 2, pp. 85-135, *National Bureau of Economic Research*, New York.

Corbridge, S. and J. Harriss, (2000): "Reinventing India: Liberalization, Hindu Nationalism and Popular Democracy", Cambridge: Polity Press.

Crosson, P.R. and N.J. Rosenberg (1989): "Strategies for Agriculture", *Scientific American,* September, pp. 78-85.

CSO (2012): "Quick Estimates of National Income, Consumption Expenditure, Saving and Capital Formation 2010-11", Central Statistics Office (CSO), Ministry of Statistics and Programme Implementation, Government of India, New Delhi, 31st January.

Dantwala, M.L. (1973): "From Stagnation to Growth: Relative Roles of Technology, Economic Policy and Agrarian Institutions," in R.T. Shand (editor): Technical Change In Asian Agriculture. Australian National University Press, Canberra, pp. 259-281.

Dantwala, M.L. (1986): "Indian Agricultural Development since Independence", Oxford and IBH Publishing Co., New Delhi, India.

Datt, G. and M. Ravallion (2009): "Has India's Economic Growth Become More Pro-poor in the Wake of Economic Reforms?", World Bank, Policy Research Working Paper 5103: Washington, D.C.

Datta, K.K., C. de Joy and M.T. Rajashekharappa (2004): "Implication of Land Degradation on Crop Productivity: Some Evidences from Saline Areas in North-West India", *Indian Journal of Agricultural Economics*, Vol. 59(1), January-March; pp. 150-163.

Debreu, G. (1951): "The Coefficient of Resource Utilisation", *Econometrica*, 19; pp. 273-292.

den Biggelaar, Christoffel, Rattan Lal, Keith Wiebe and Vince Breneman (2004): "The Global Impact of Soil Erosion on Productivity (I): Absolute and Relative Erosion-Induced Yield Losses", *Advances in Agronomy*; 81; pp. 1-48.

Desai, Bhupat M. (1994): "Contribution of Institutional Credit, Self-finance and Technological Change to Agricultural Growth in India", *Indian Journal of Agricultural Economics*.

Desai, Bhupat M., Errol D'Souza, John W Mellor, Vijay Paul Sharma and Prabhakar Tamboli (2011): "Agricultural Policy Strategy, Instruments and Implementation: A Review and the Road Ahead", *Economic and Political Weekly*, December 31; XLVI (53); pp. 42-50.

Desai, Bhupat, M. and N. V. Namboodiri (1997): "Determinants of Total Factor Productivity in Indian Agriculture", *Economic and Political Weekly*, December 27, pp. A165-A171.

Deshpande R.S., M.J. Bhende and P. Thippaiah (2004): "Crops and Cultivation, State of Indian Farmer: A Millennium Study", Vol. 9, Academic Foundation, New Delhi.

Dev, S. Mahendra and Ajit Ranade (1998): "Rising Food Prices and Rural Poverty: Going Beyond Correlations", *Economic and Political Weekly*, 33 (39); pp. 2529-2536.

Dev, Mahendra S. and Rao N.C. (2010): "Agricultural Price Policy, Farm Profitability and Food Security", *Economic and Political Weekly*, 45(26 & 27); pp. 174-182.

Dhaliwal, G.S. and R. Arora (1996): "Principles of Insect Management", Commonwealth Publishers, New Delhi.

Dhawan, B.D. (1988): "Irrigation in India's Agricultural Development: Productivity, Stability and Equity", Sage Publications, New Delhi.

Dhawan, B.D. (1995): "Groundwater Depletion, Land Degradation and Irrigated Agriculture in India", Commonwealth Publishers; New Delhi.

Dholakia, Ravindra H. and Bakul H. Dholakia (1993): "Growth of Total Factor Productivity in Indian Agriculture", *Indian Economic Review*, XXVIII (1); pp. 25-40.

Diewart, E.W. and A.O. Nakamura (2003): "A Survey of Empirical Methods of Productivity Measurement and Analysis", in J.J. Heckman and E.E. Leamer (eds.), "Handbook of Econometrics", Vol. 6, Elsevier Publication, North Holland.

Diewart, W.E. (1976): "Exact and Superlative Index Numbers", *Journal of Econometrics*, 4(2); pp. 116-145.

Diewart, W.E. (1981): "The Economic Theory of Index Numbers: A Survey", in A. Deaton (ed.), "Essays in the Theory and Measurement of Consumer Behaviour in Honour of Sir Richard Stone", Cambridge University Press, London.

Dobermann, A., P.C. Santa Cruz and K.G. Cassman (1995): "Potassium Balances and Soil Potassium Supplying Power in Intensive Irrigated Rice Ecosystems," in 24th Colloquium of IPI, Chiang Mai, Thailand; pp. 199-229.

Domar, E. (1961): "On the Measurement of Technological Change", *The Economic Journal,* Vol. 71; pp. 709-729.

Durbin, J. and S.J. Koopman (2001): "Time Series Analysis by State Space Methods", Oxford University Press.

Easter, William K., Martin E. Abel, and George Norton (1976): "Regional Differences in Agricultural Productivity in Selected Areas of India", Staff Paper P 76-12, Department of Agricultural and Applied Economics, University of Minnesota.

Easterly, W. and R. Levine (2001): "What Have We Learned from a Decade of Empirical Research on Growth? It is Not Factor

Accumulation: Stylized Facts and Growth Models", *The World Bank Economic Review*, 15(2); pp. 177-219.

Echevarria, C. (1997): "Changes in Sectoral Composition Associated With Economic Growth", *International Economic Review,* 38(2); pp. 431-452.

Ellis, F. (2000): "Rural Livelihoods Diversity in Developing Countries", Oxford University Press, Oxford.

Environment Statistics of Punjab (2011), Economic and Statistical Organisation, Punjab.

Erenstein, Olaf (2009): "Leaving the Plow Behind: Zero-Tillage Cultivation in the Indo-Gangetic Plains", in David J. Spielman and Rajul Pandya-Lorch, Millions Fed (eds.), Proven Successes in Agricultural Development, Washington D.C.: International Food Policy Research Group.

Esposti, R. (2000): "Stochastic Technical Change and Procyclical TFP: The Italian Agriculture Case", *Journal of Productivity Analysis*, 14; pp. 119-141.

Evans, L.T. (1993): "Crop Evolution, Adaptation and Yield", Cambridge University Press, Cambridge.

Evenson, R.E. and Carl E. Pray and Mark W. Rosegrant (1999): "Agricultural Research and Productivity Growth in India", Research Report 109; International Food Policy Research Institute, Washington, D.C.

Evenson, R.E. (1985): "Output Supply and Input Demand Effects of High Yielding Rice and Wheat Varieties in North Indian Agriculture", *Indian Journal of Quantitative Economics*, I (1); pp. 85-96.

Evenson, R.E., and D. Gollin (2003): "Assessing the Impact of the Green Revolution, 1960 to 2000", *Science*, 300; pp. 758-762.

Fabricant, Solomon (1959): "Basic Facts of Productivity Growth", *National Bureau of Economic Research*, Occasional Paper No. 63.

Fabricant, Solomon (1961): "Introduction", in John, W. Kendrick's, Productivity Trends in US, Princeton University Press, Princeton.

Fan, Shenggen (1991): "Effects of Technological Change and Institutional Reform on Production Growth in Chinese Agriculture", *American Journal of Agricultural Economics*, 72; pp. 266-275.

FAO (2001): "Environmental Impacts of Productivity: Enhancing Crop Research—A Critical Review", A Report from TAC's Standing Panel on Impact Assessment, TAC Secretariat, UNO.

Färe, Rolf (1988): "Fundamentals of Production Theory", Lecture Notes in Economics and Mathematical Systems, Springer-Verlag, Heidelberg.

Färe, Rolf, Shawna Grosskopf and C.A.K. Lovell (1994, a): "Production Frontiers", Cambridge University Press, Cambridge.

Färe, Rolf, Shawna Grosskopf, M. Norris and Z. Zhang (1994, b): "Productivity Growth, Technical Progress and Efficiency Change in Industrial Countries", *American Economic Review*, pp. 66-89.

Farrell, M. (1957): "The Measurement of Productive Efficiency", *Journal of Royal Statistical Society*, Series A, CXX (3); pp. 253-290.

Feder, G., Lau, L.J. and R.H. Slade (1987): "Does Agricultural Extension Pay? The Training and Visit System in Northwest India", *American Journal of Agricultural Economics,* 69; pp. 677-686.

Felipe, J. and F. Fisher (2003): "Aggregation in Production Functions: What Applied Economists Should Know", *Metroeconomica*, 54; pp. 208-262.

Felipe, J. and J.S.L. McCombie (2004): "To Measure or Not To Measure TFP Growth? A Reply to Mahadevan", *Oxford Development Studies*, 32(2); pp. 321-327.

Flinn, J.C. and S.K. De Datta (1984): "Trends in Irrigated Rice Yields under Intensive Cropping At Philippines Research Stations", *Field Crops Research*, 5; pp. 201-216.

Førsund, F. R. and Nikias Sarafogulu (2000): "On the origins of DEA", Memorandum 24, Department of Economics, University of Oslo.

Førsund, F.R., C.A.K. Lovell and P. Shimdt (1980): "A Survey of Frontier Production Functions and Their Relationship to Efficiency Measurement", *Journal of Econometrics*, 13; pp. 5-25.

Foster S. and L. Candela (2008): "Diffuse Groundwater Quality Impacts from Agricultural Land-Use: Management and Policy Implications of Scientific Realities", in Groundwater Science and Policy: An International Overview, RSC Publishing, London, UK, pp. 454-470.

Foster, A.D. and M.R. Rosenzweig (1996): "Technical Change and Human Capital Returns and Investments: Evidence from the Green Revolution", *American Economic Review,* 86(4); pp. 931-953.

Fuglie, K.O. (2010): "Total Factor Productivity in the Global Agricultural Economy: Evidence from FAO Data", (Chapter 4) in Julian Alston, Bruce Babcock and Philip Pardey (eds.), The Shifting Patterns of Agricultural Production and Productivity Worldwide, Midwest Agribusiness Trade and Research Information Centre

(MATRIC): Iowa State University, Ames, Iowa.

Gandhi, Vasant P. and N.V. Namboodiri (2009): "Groundwater Irrigation in India: Gains, Costs and Risks", Working Paper 2009-03-08, March, Indian Institute of Management, Ahmedabad, India.

Gandhi, Vasant P. and N.V. Namboodiri (2010): "Economics of Bt Cotton vis-à-vis Non-Bt Cotton in the State of Maharashtra, India", Allied Publishers Pvt. Ltd., New Delhi.

Gemmell, N., T. Lloyd, and M. Mathew (2000): "Agricultural Growth and Intersectoral Linkages in Developing Economy", *Journal of Agricultural Economics*, 51(3); pp. 353-370.

Ghosh, Jayati (2005): "Trade Liberalisation in Agriculture: An Examination of Impact and Policy Strategies with Special Reference to India", Background Paper for Human Development Report 2005, New Delhi.

Gollin, D., S.L. Parente and R. Rogerson (2002): "The Role of Agriculture in Development", *American Economic Review*, 92(2); pp. 160-164.

Government of India (GoI) (2002): "India Vision 2020", Planning Commission, December, New Delhi.

Government of India (GoI) (2005): "Agriculture and Food Security", Mid-term Review of the Tenth Five Year Plan, Part II, Chapter 5, Planning Commission, New Delhi.

Government of India (GoI) (2011): "Faster, Sustainable and More Inclusive Growth: An Approach to the 12th Five Year Plan (2012-2017)", *Planning Commission*, New Delhi: Government of India.

Government of India (GoI) (Various Issues): "Agricultural Statistics at Glance", Directorate of Economics and Statistics, Department of Agriculture and Cooperation, Ministry of Agriculture, New Delhi.

Government of India (GoI): "Comprehensive Scheme for the Study of Cost of Cultivation of Principal Crops in India", various volumes.

Government of Punjab (2007): "Punjab: The Dance of Water Table", The Punjab State Farmers Commission, Mohali.

Greene, William H. (1980a): "Maximum Likelihood Estimation of Econometric Frontier Functions", *Journal of Econometrics,* 13; pp. 27-56.

Greene, William H. (1980b): "On the Estimation of a Flexible Frontier Production Model", *Journal of Econometrics,* 13; pp. 101-105.

Griliches, Zvi (1963): "Estimates of the Aggregate Agricultural Production Function from Cross-Sectional Data", *Journal of Farm Economics*, 45(2); pp. 419-428.

Griliches, Zvi (1994): "Productivity, R&D, and the Data Constraint", *American Economic* Review, 84(1), March; pp. 1-23.

Griliches, Zvi (1995): "The Discovery of the Residual", NBER Working Paper No. 5348.

Griliches, Zvi (1996): "The Discovery of the Residual: A Historical Note", *Journal of Economic Literature*, 34, September, pp. 1324-1330.

Grosskopf, S. (1993): "Efficiency and Productivity", Chapter 4, in H. Fried, C.A.K. Lovell and S. Schmidt (eds.), The Measurement of Productive Efficiency: Techniques and Applications, Oxford.

Guilkey, D.K., C. Lovell and R.C. Sickles (1983): "A Comparison of The Performance of Three Flexible Functional Forms", *International Economic Review,* 24.

Gulati, Ashok (2009): "Emerging Trends in Indian Agriculture: What Can We Learn from These?",2nd Prof. Dayanatha Jha Memorial Lecture, National Centre for Agricultural Economics and Policy Research, New Delhi.

Gulati, Ashok and A. Sharma (1997): "Freeing Trade in Agriculture: Implications for Resource Use Efficiency and Cropping Pattern Changes", *Economic and Political Weekly*, December 17, 32(52); pp. A154-A164.

Gulati, Ashok, Surbhi Jain and Nidhi Satija (2013): "Rising Farm Wages in India: The 'Pull' and 'Push' Factors", Discussion Paper No. 5, April, Commission for Agricultural Costs and Prices; New Delhi, Government of India, accessed on 15th November, 2013.

Gumbau-Albert, Mercedes (1998): "Regional Technical Efficiency: A Stochastic Frontier Approach," *Applied Economics Letters*, 5, pp. 623-726.

Gumbau-Albert, Mercedes (2000): "Efficiency and Technical Progress: Sources of Convergence in the Spanish Regions", *Applied Economics*, 32; pp. 467-478.

Gupta, Anil K. (2004): "Origin of Agriculture and Domestication of Plants and Animals Linked to Early Holocene Climate Amelioration", *Current Science*, 87(1), July, pp. 54-59.

Gupta, R.K. and I.P. Abrol, (2000): "Salinity Build-Up and Changes in the Rice-Wheat System of the Indo-Gangetic Plains", *Experimental Agriculture,* April, 36(2); pp. 273-284.

Hall, Robert E. and Charles I. Jones (1999): "Why Do Some Countries Produce So Much Output per Worker than Others", *Quarterly Journal of Economics*, 114, pp. 83-116.

Hanoch, G. and M. Rothschild (1972): "Testing the Assumptions of Production Theory: A Non-parametric Analysis", *Journal of Political Economy*, 80; pp. 256-275.

Harberger, Arnold C. (1998): "A Vision of the Growth Process", *The American Economic Review*, 88(1), pp. 1-32.

Harrington, L. (2001): "Synthesis of Systems Diagnosis: Is the Sustainability of the Rice-Wheat Cropping System Threatened?: An Epilogue"; pp. 119-132, in Kataki, P.K., (ed.), The Rice-Wheat Cropping System of Asia: Trends, Constraints, Productivity and Policy, Food Products Press, New York, USA.

Harvey, A. and S. Wren-Lewis (1986): "Stochastic Trends in Dynamic Regression Models: An Application to the Employment-Output Equations", *Economic Journal*, 96(384), December; pp. 975-985.

Hastie, Trevor and Robert Tibshirani (1986): "Generalized Additive Models", *Statistical Science*; Vol. 1, No. 3, pp. 297-318.

Hayami Y. and V.W. Ruttan (1970): "Agricultural Productivity Differences among Countries", *American Economic Review*, 60(5); pp. 895-911.

Hayami Y. and V.W. Ruttan (1985): "Agricultural Development: An International Perspective", Baltimore: The Johns Hopkins University Press.

Hazell, B.R. Peter, V.N. Misra and Behjat Hojjati (1999): "Role of Terms of Trade in Indian Agricultural Growth: A National and State Level Analysis", EPTD Discussion Paper 15, International Food Policy Research Institute, Washington D.C.

Hesmati, A. (1996): "On Single and Multiple Time-Trends Representations of Technical Change", *Applied Economic Letters*, Vol. 3; pp. 495-499.

Himanshu, Peter Lanjouw, Abhiroop Mukhopadhyay and Rinku Murgai (2011): "Non-farm Diversification and Rural Poverty Decline: A Perspective from Indian Sample Survey and Village Study", Working Paper 44, Asia Research Centre, London School of Economics and Political Science: London.

Hjalmarsson, L., S.C. Kumbhakar and A. Heshmati (1996): "DEA, DFA and SFA: A Comparison", *Journal of Productivity Analysis*, 7; pp. 303-327.

Hobbs, Peter R. and Raj K. Gupta (2003): "Rice-Wheat Cropping Systems in the Indo-Gangetic Plains: Issues of Water Productivity in Relation to New Resource Conserving Technologies", in J.W. Kijne, R. Barker and D. Molden (eds.), Water Productivity in

Agriculture: Limits and Opportunities for Improvement, CAB International, Wallingford, UK, in association with the International Water Management Institute (IWMI), Sri Lanka, 2004.

Hoch, I. (1962): "Estimation of Production Function Parameters Combining Time-Series and Cross-Section Data", *Econometrica,* 30; pp. 34-53.

Huang, C. and J. Liu (1994): "Estimation of a Non-Neutral Stochastic Frontier Production Function", *Journal of Productivity Analysis*, 5; pp. 171-180.

Huang, C. J. and F.S. Bagi (1984): "Technical Efficiency on Individual Farms in Northwest India", *Southern Economic Journal*, 51; pp. 108-115.

Huang, Jikun and Scott Rozelle (1995): "Environmental Stress and Grain Yields in China", *American Journal of Agricultural Economics*, November, 77, pp. 853-864.

Hulten, Charles R. (2000): "Total Factor Productivity: A Short Biography", in Charles R. Hulten, Edwin R. Dean, and Michael J. Harper (eds.), New Directions in Productivity Analysis, Studies in Income and Wealth, The University of Chicago Press for the National Bureau of Economics Research, Chicago.

Humphries, H. and S. Knowles (1998): "Does Agriculture Contribute To Economic Growth? Some Empirical Evidence", *Applied Economics*, 30(6); pp. 775-781.

ICAR, Indian Council of Agricultural Research (2010): "Degraded and Wasteland of India: Status and Spatial Distribution", New Delhi.

Islam N. (1995): "Growth Empirics: A Panel Data Approach", *Quarterly Journal of Economics*, 4; pp. 1127-1170.

Hausman, J.A. (1978): "Specification Tests in Econometrics", *Econometrica*; 46; pp. 1251-1271.

IWMI (International Water Management Institute) (2010): "Influencing Irrigation Policy in India: Success Stories", Issue 6, p. 2, Colombo, Sri Lanka: International Water Management Institute.

Janaiah, A., K. Otsuka and M. Hossain (2005): "Is The Productivity Impact of the Revolution in Rice Vanishing?", *Economic and Political Weekly*, 40(53); pp. 5596-5600.

Janaiah, Aldas and Lalith Achoth (2006): "Is the Green Revolution Vanishing? Empirical Evidence from TFP Analysis for Rice", Poster paper prepared for presentation at the International Association of Agricultural Economists Conference, Gold Coast, Australia, August 12-18.

Jenkins, R. (1999): "Democratic Politics and Economic Reforms in India", Cambridge University Press, Cambridge: U.K.

Jha, Dayanath and Robert E. Evenson (1973): "The Contribution of Agricultural System to Agricultural Production in India", *Indian Journal of Agricultural Economics*, 52(3); pp. 96-110.

Jha, Raghbendra and Mark J. Rhodes (1999): "Some Imperatives of the Green Revolution: Technical Efficiency and Ownership of Inputs in Indian Agriculture", *Agricultural and Resource Economics Review*, 28(1); pp. 57-64.

Jorgenson (1995): "Productivity: Post-War U.S. Economic Growth", Volume 1, Cambridge: MIT Press.

Joshi, P.K. and N.K. Tyagi (1991): "Sustainability of Existing Farming System in Punjab and Haryana: Some Issues in Groundwater Use", *Indian Journal of Agricultural Economics*, 46(3); pp. 412-421.

Joshi, P.K. (1987): "Effects of Surface Irrigation Land Degradation: Problems and Strategies", *Indian Journal of Agricultural Economics*, 42; pp. 416-423.

Joshi, P.K. and D. Jha (1991): "Farm-level effects of Soil Degradation", in Sharda Sahayak Irrigation Project, Working Paper on Future Growth in Indian Agriculture No. 1 IFPRI, Washington, D.C.

Junankar, P.N. (1980): "Tests of the Profit Maximizing Hypothesis: A Study of Indian Agriculture", *Journal of Development Studies*,16; pp. 186-203.

Kalirajan, K.P. (1981): "An Econometric Analysis of Yield Variability in Paddy Production", *Canadian Journal of Agricultural Economics*, 29, pp. 283-294.

Kalirajan, K.P. and R.T. Shand (1985): "Types of Education and Agricultural Productivity: a Quantitative Analysis of Tamil Nadu Rice Farming", *Journal of Development Studies*, 21; pp. 232-243.

Kalirajan, K.P. and R.T. Shand (1989): "A Generalised Measure of Technical Efficiency", *Applied Economics*, 21; pp. 225-234.

Kalirajan, K.P. and R.T. Shand (1997): "Sources of Output Growth in Indian Agriculture", *Indian Journal of Agricultural Economics*, 52(4); pp. 693-706.

Kallianpur, R. and R. Diwan (1985): "Productivity Growth: Scale and Technological Analysis of Punjab Wheat Farms", *Indian Journal of Quantitative Economics*, I(1); pp. 61-84.

Kanwar, J.S. and J.C. Katyal (eds.) (1997): "Plant Nutrients Needs, Supply, Efficiency and Policy Issues: 2000-2025", NAAS, New Delhi.

Kerr, John M (1996): "Sustainable Development of Rainfed Agriculture in India", EPTD Discussion Paper No. 20, IFPRI, Washington.

Khush, Gurdv (2003): "Productivity Improvements in Rice", *Nutrition Reviews*, June, 61(6); pp. S114-S116.

Kislev Y. and W. Peterson (1996): "Economies of Scale in Agriculture: A Re-examination of Evidence", in J.M. Antle and D.A. Summer, (eds.), The Economics of Agriculture: Papers in Honor of D. Gale Johnson, Volume 2; pp. 156-170, The University of Chicago Press.

Kodde, D.A. and F.C. Palm (1986): "Wald Criteria for Jointly Testing Equality and Inequality Restrictions", *Econometrica*, 54; pp.1243-1248.

Kogel, T. and A. Prskawetz (2001): "Agricultural Productivity Growth and Escape from the Malthusian Trap", *Journal of Economic Growth*, 6; pp. 337-357.

Kohli, U. (1990): "Growth Accounting in the Open Economy: Parametric and Nonparametric Estimates", *Journal of Economic and Social Measurement*, 16, pp. 125-136.

Koop, G. and M.F.J. Steel (2001): "Bayesian Analysis of Stochastic Frontier Models", in Baltagi, B.H. (Ed.), A Companion to Theoretical Econometrics, Oxford, Blackwell; pp. 520- 537.

Koopmans, T. (1951): "An Analysis of Production as an Efficient Combination of Activities", in T. Koopmans (ed.), Activity Analysis of Production and Allocation, John Wiley and Sons, New York; Cowles Commission for Research in Economics, Monograph No. 13.

Krishnaji, N. (1990): "Agricultural Price Policy: A Survey with Reference to Indian Foodgrain Economy", *Economic and Political Weekly*, 25(26); pp. A54-A63.

Parikh, Kirit S., A. Ganesh Kumar and Gangadhar Darbha (2003): "Growth and Welfare Consequences of Rise in MSP", *Economic and Political Weekly*, 38(9), pp. 891-895.

Krugman, Paul, (1996): "The Myth of Asia's Miracle" in *Pop Internationalism,* MIT Press: Cambridge.

Kulshrestha, A.C. (2004): "Agricultural Statistics: Data Availability, Requirements and Gaps from the Perspective of NAS", *Journal of Indian Society of Agricultural Statistics*, 57 (Special Volume); pp. 345-368.

Kumar, A. and D.S. Yadav (2001): "Long-Term Effects of Fertilizers on the Soil Fertility and Productivity of a Rice-Wheat System", *Journal of Agronomy Crop Science*, 186; pp. 47-54.

Kumar, Praduman (2001): "Agricultural Performance and Productivity" in S.S. Acharya and D.P. Chaudhry (eds.), Indian Agricultural Policy at the Crossroads: Priorities and Agenda, Rawat Publications: Jaipur and New Delhi.

Kumar, Praduman and Mruthyunjaya (1992): "Measurement and Analysis of Total Factor Productivity in Wheat", *Indian Journal of Agricultural Economics*, 47(1); pp. 451-458.

Kumar, Praduman, M.W. Rosegrant and P. Hazell (1995): "Cereal Prospects in India to 2020: Implications for Policy", IFPRI 2020 Policy Brief No. 23; Washington, D.C.: IFPRI.

Kumar, Praduman, A. Kumar and Surabhi Mittal (2004b): "Total Factor Productivity of Crop Sector in the Indo-Gangetic Plain of India: Sustainability Issues Revisited", *Indian Economic Review*, 39(1); pp. 169-201.

Kumar Praduman and Surabhi Mittal (2006): "Agricultural Productivity Trends in India: Sustainability Issues", *Agricultural Economics Research Review*, Vol. 19, pp. 71-88.

Kumbhakar, Subal C. (1990): "Production Frontiers, Panel Data, and Time-Varying Technical Inefficiency", *Journal of Econometrics*, 46, October; pp. 201-211.

Kumbhakar, Subal C. and C.A. Knox Lovell (2000): "Stochastic Frontier Analysis", Cambridge University Press, Cambridge.

Ladha, J.K., K.S. Fischer, M. Hossain, P.R. Hobbs, and B. Hardy (eds.) (2000): "Improving the Productivity and Sustainability of Rice-Wheat Systems of the Indo-Gangetic Plains: A Synthesis of NARS-IRRI Partnership Research", Discussion Paper No. 40, International Rice Research Institute (IRRI).

Ladha, J.K., D. Dawe, H. Pathak, A.T. Padre, R.L. Yadav, B. Singh, Y. Singh, Y. Singh, P. Singh, A.L. Kundu, R. Sakal, N. Ram, A.P. Regmi, S.K. Gami, A.L. Bhandari, R. Amin, C.R. Yadav, E.M. Bhattarai, S. Das, H.P. Aggarwal, R.K. Gupta, and P.R. Hobbs (2003): "How Extensive are Yield Declines in Long-term Rice: Wheat Experiments in Asia?", *Field Crops Research*, 81; pp. 159-180.

Lall, Sanjaya (1993): "Understanding Technology Development", *Development and Change*, 24; pp. 719-753.

Lansink A., E. Silva and S. Stefanou (2000): "Decomposing Productivity Growth Allowing Efficiency Gains and Price Induced Technical Progress", *European Review of Agricultural Economics*, 27(4); pp. 497-518.

Lau, L. and P. Yotopoulas (1961): "A Test of Relative Efficiency and Applications to Indian Agriculture", *American Economic Review*, 66; pp. 94-110.

Law, M.T. (2000): "Productivity and Economic Performance: An Overview of the Issues", *Public Policy Sources*, No. 37.

Lin, X., C. Yin and D. Xu (1996): "Input and Output of Soil Nutrients in High-yield Paddy Fields in South China", in Proceedings of the International Symposium on Maximizing Rice Yields through Improved Soil and Environmental Management, Khon Kaen, Thailand; pp. 93-97.

Lindert, Peter H. (2000): "Shifting Ground: The Changing Agricultural Soils of China and Indonesia", Cambridge, Massachusetts: MIT Press.

Lipton, M. (2006): "Can Small Farmers Survive, Prosper, or be the Key Channel to Cut Mass Poverty?", *Journal of Agriculture and Development Economics*, Vol. 3(1); pp. 58-85.

Lynam, J.K. and R.W. Herdt (1989): "Sense and Sustainability: Sustainability as an Objective in International Agricultural Research", *Agricultural Economics*; 3; pp. 381-398.

Mahadevan, R. (2003): "To Measure or Not to Measure TFP Growth?", *Oxford Development Studies*, 31(3); pp. 365-378.

Mahadevan, R. (2004): "To Measure or Not to Measure TFP Growth? Comment on "Unequivocal No" Reply", *Oxford Development Studies*, 32 (2); pp. 329-330.

Malmquist, S. (1953): "Index Numbers and Indifference Curves", *Trabajos de Estatistica*, Vol. 4, pp. 34-51.

Mas-Colell, M.D. Whiston and J.R. Green (1995): "Microeconomic Theory", Oxford University Press, New Delhi.

Meeusen, W. and J. van den Broeck (1977): "Efficiency Estimation from Cobb-Douglas Production Functions with Composed Errors", *International Economic Review*, 18; pp. 433-444.

Millet, Tracey and George Tolley (1989): "Technological Adoption and Agricultural Price Policy", *American Journal of Agricultural Economics*, 71; pp. 847-857.

Misra, V.N. and P.R. Hazell (1996): "Terms of Trade, Rural Poverty, Technology and Investment: The Indian Experience: 1952-53 to 1990-91", *Economic and Political Weekly*, 31(13), March 30; pp. A2-A13.

Mody, Ashoka and Krishna Srinivasan (1998): "Japanese and U.S. Firms as Foreign Investors: Do They March to the Same Tune?",

Canadian Journal of Economics, October, 31(4): pp. 778-799.

Moench, Marcus (2003): "Groundwater and Food Security in India" in Prasad, Kamta (ed.), Water Resources and Sustainable Development: Challenges of 21st Century, Shipra Publications, Delhi; pp. 148-167.

Mooij, J. (2005): "Introduction", in J. Mooij (ed.), The Politics of Economic Reforms In India, Sage Publications: New Delhi.

Moroney, John R. and C.A.K. Lovell (1997): "The Relative Efficiencies of Market and Planned Economies", *Southern Economic Journal*, 63; pp. 1084-1093.

Motsara, M.R. (2002): "Available Nitrogen, Phosphorus, and Potassium Status of Indian Soil as Depicted by Soil Fertility Maps", *Fertilizer News*, Vol. 47(8); pp. 15-21.

Mueller, Nathaniel D., James S. Gerber, Matt Johnston, Deepak K. Ray, Navin Ramankutty and Jonathan A. Foley (2012): "Closing Yield Gaps through Nutrient and Water Management", *Nature*, 11 October, Vol. 490; pp. 254-257.

Mujeri, Mustafa K., Siban Shahana, Tahreen Tahrima Chowdhury and Khondoker Tanveer Haider (2012): "Improving the Effectiveness, Efficiency and Sustainability of Fertilizer Use in South Asia", The Global Development Network (GDN), June.

Muller, J. (1974): "On Sources of Measured Technical Efficiency: The Impacts of Information", *American Journal of Agricultural Economics*, 56; pp. 730-738.

Mundlak, Yair (2000): "Agriculture and Economic Growth: Theory and Measurement", Harvard University Press: Cambridge.

Mundlak, Yair, Donald F. Larson and Rita Butzer (1999): "Rethinking Within and Between Regressions: The Case of Agricultural Production Functions", *Annales D'économie Et De Statistique*, 55-56; pp. 475-501.

Munshi, K. (2004): "Social Learning in a Heterogenous Population: Technology Diffusion in Indian Green Revolution", *Journal of Development Economics*, 73(1); pp. 185-213.

Murgai, Rinku, Mubarak Ali and Derek Byerlee (2001): "Productivity Growth and Sustainability in Post-Green Revolution Agriculture: The Case of the Indian and Pakistan Punjabs", *The World Bank Research Observer*, 16(2); pp. 199-218.

Mutert, E. (1996): "Plant Nutrient Balance in Asia and the Pacific Region: Facts and Consequences for Agricultural Production", in APO-FFTC Seminar on Appropriate Use of Fertilizers; pp. 73-112.

National Agriculture Policy, July 2000.

Navroz, K. Dubash (2002): "Tubewell Capitalism: Groundwater Development and Agrarian Change in Gujarat", Oxford University Press, New Delhi.

Nayyar, Deepak and Abhijit Sen (1994): "International Trade and the Agricultural Sector in India", *Economic and Political Weekly*, May 14; pp. 1187-1203.

Nin-Pratt and Fan Yu (2010): "Comparisons of Agricultural Productivity Growth in China and India", *Journal of Productivity Analysis,* 33; pp. 209-223.

Nishimizu, M. and J.M. Page (1982): "Total Factor Productivity Growth, Technical Progress and Technical Efficiency Change: Dimensions of Productivity Change in Yugoslavia, 1965-78", *Economic Journal*, 92; pp. 920-936.

Øksendal, B. (1985): "Stochastic Differential Equations: An Introduction with Applications", Springer-Verlag, Berlin.

Olsson, O. and D.A. Hibbs (2005): "Biogeography and Long-Run Economic Development", *European Economic Review*, 49(4); pp. 909-938.

Ostrom E. (1990): "Governing the Commons: The Evolution of Institutions for Collective Action", Cambridge University Press, Cambridge, UK.

Pagiola, Stefano (1995b): "Environmental and Natural Resource Degradation in Intensive Agriculture in Bangladesh", Environment Department Paper No. 15, Washington D.C.: The World Bank.

Patnaik, Utsa (1999): "The Long Transition: Essays on Political Economy", Tulika: New Delhi.

Paul, S. and R. Mehta (1991): "Technology, Factor Demand and Substitution in Indian Agriculture: 1960-83", *Indian Journal of Agricultural Economics*, Vol. 46; pp. 530-543.

Piesse, Jenifer and Colin Thirtle (2000): "A Stochastic Frontier Approach to Firm Level Efficiency, Technological Change, and Productivity during the Early Transition in Hungary", *Journal of Comparative Economics*, 28; pp. 473-501.

Pillai, Renuka (2001): "An Analysis of Paddy Productivity Growth in West Bengal and Orissa", *Indian Journal of Agricultural Economics*, 56(4); October-December; pp. 613-630.

Pimentel, David, C. Harvey, P. Resosudarmo, K. Sinclair, D. Kurz, M. McNair, S. Crist, L. Shpritz, L. Fitton, R. Saffouri and R. Blair (1995): "Environmental and Economic Costs of Soil Erosion and

Conservation Benefits", *Science*, February, 267; pp. 1117-1123.

Pingali, P.L. and M.W. Rosegrant (1994): "Confronting the Environmental Consequences of the Green Revolution", in Asia. EPTD Discussion Paper No. 2, August, IFPRI, Washington, D.C.

Pingali P.L., M. Hossain and R. V. Gerpacio (1997): "Asian Rice Bowls: The Returning Crisis?", Wallingford (UK): CAB International, in Association with the International Rice Research Institute; p. 341.

Pingali, P.L. and H.P. Binswanger (1988): "Population Density and Farming Systems: The Changing Locus of Innovations and Technical Change", in Ronald D. Lee, W. Brian Arthur, Allen C. Kelley, Gerry Rodgers and T. N. Srinivasan (eds.), Population, Food and Rural Development, International Studies in Demography: Oxford: Clarendon Press.

Pingali P.L. and M.W. Rosegrant (2001): "Intensive Food Systems in Asia: Can the Degradation Problems be Reversed?", in D.R. Lee and C.B. Barrett CB (eds.), Trade-offs or Synergies? Agricultural Intensification, Economic Development and Environment; Wallingford (UK): CABI.

Pinstrup-Andersen, P. and R. Pandya-Lorch (1994): "Alleviating Poverty, Intensifying Agriculture and Effectively Managing Natural Resources", Food, Agriculture and the Environment Discussion Paper 1, IFPRI, Washington D.C.

Postel, S. (1993): "Water and Agriculture", in Gleick P.H. (ed.), Water in Crisis: A Guide to the World's Fresh Water Resources, Oxford University Press: New York, N.Y.

Postel, S. (1989): "Water for Agriculture: Facing the Limits", Worldwatch Paper 93, December.

Pretty, J. and R. Hine (2005): "Pesticide Use and the Environment", in J.N. Pretty (ed.), The Pesticide Detox: Towards a More Sustainable Agriculture, London: Earthscan; pp. 1-22.

Rada, Nicholas E. (2013): "Agricultural Growth in India: Examining the Post-Green Revolution Transition", 2013 Annual Meeting, August 2013, Agricultural and Applied Economics Association, Washington, D.C.

Raghavan, M. (2008): "Changing Pattern of Input Use and Cost of Cultivation Use", *Economic and Political Weekly*, June 28; pp. 123-129.

Ram, S. and G.C. Sharma (1997-98): "Characteristics of Rice-Wheat Production System in Western Pradesh", *Journal of Farming Systems Research and Development*, Vols. 3 and 4, Nos. 1 and 2.

Ramakrishna, G. (1993): "Growth and Fluctuations in Indian Agriculture", *Asian Economic Review*, The Journal of the Indian Institute of Economics, (35), April, pp. 47-60.

Ramakumar, R. (2010): "Continuity and Change: Notes on Agriculture in New India", in Anthony D'Costa (ed.), A New India? Critical Perspectives in the Long Twentieth Century, London: Anthem Press.

Ranis, Gustav (2004): "Arthur Lewis' Contribution to Development Thinking and Policy", *The Manchester School*, (72) 6; pp. 712-723.

Rao, C.H. Hanumantha (1994): "Agricultural Growth, Rural Poverty, and Environmental Degradation in India", Oxford University Press, New Delhi.

Rao, C.H. Hanumantha (2004): "Preface to the Millennium Study", in State of the Indian Farmer: A Millennium Study, New Delhi: Academic Foundation, on behalf of the Department of Agriculture and Cooperation, Ministry of Agriculture, Government of India.

Rao, C.H. Hanumantha and Ashok Gulati (1994): "Indian Agriculture: Emerging Perspectives and Policy Issues", New Delhi: ICAR and Washington, D.C.: IFPRI.

Rao, E.V.S. Prakasa and K. Puttanna (2000): "Nitrates, Agriculture and Environment", *Current Science*, Vol. 79(9); 10th November, pp. 1163-1168.

Rao, Mohan J. and Servaas Storm (1998): "Distribution and Growth in Indian Agriculture" in T.J. Byres (ed.), The Indian Economy: Major Debates Since Independence, Oxford University Press, New Delhi; pp. 193-248.

Rao, N. Chandrasekhara and S. Mahendra Dev (2010): "Agricultural Price Policy, Farm Profitability and Food Security", *Economic and Political Weekly*, XLV(26-27), June 26; pp.174-182.

Rao, V.M. (2001): "The Making of Agricultural Price Policy: A Review of CACP Reports", *Journal of Indian School of Political Weekly*, 13 (1); pp. 1-28.

Sen, Abhijit (1999): "Agricultural Price Policy: Achievements and the Emerging Problems", in Y.V. Krishna Rao (ed.), New Challenges Facing Indian Agriculture, Hyderabad: Visalandhra Publishing House.

Rawal, Vikas (2001): "Agrarian Reform and Land Markets: A Study of Land Transactions in Two Villages of West Bengal, 1977-1995", *Economic Development and Cultural Change*, 49, April; pp. 611-629.

Ray, Subhash C. (1985): "Measurement and Test of Efficiency of Farms in Linear Programming Models: A Study of West Bengal Farms", *Oxford Bulletin of Economics and Statistics*, 47, pp. 371-386.

Robinson, Joan (1953): "The Production Function and the Theory of Capital", *Review of Economic Studies*, 21(2); pp. 81-106.

Rola, A.C. and P. Pingali (1993): "Pesticides, Rice Productivity, and Farmers' Health: An Economic Assessment", IRRI-World Resources Institute, WRI, Los Baños, Laguna, Philippines.

Rosegrant, M.W. and J.A. Roumasset (1988): "Economic Feasibility of Green Manure in Rice-Based Cropping System", in Green Manure in Rice Farming, Los Banos: IRRI.

Rosegrant, M. and R.E. Evenson (1992): "Agricultural Productivity and Sources of Growth in South Asia", *American Journal of Agricultural Economics*, 73(3); pp. 757-761.

Rosegrant, M.W. and R.E. Evenson (1993): "Agricultural Productivity and Growth in Pakistan and India: A Comparative Analysis", *Pakistan Development Review*, 32(4); pp. 433-451.

Rosegrant, M.W., and P.B.R. Hazell (2000): "Transforming The Rural Asian Economy: The Unfinished Revolution", Oxford University Press: Hong Kong.

Rudra, Ashok (1982): "Indian Agricultural Economics: Myths and Realities", Allied Publishers, New Delhi.

Ruttan, Vernon W. (2002): "Productivity Growth in World Agriculture: Sources and Constraints", *Journal of Economic Perspectives*, 16; pp. 161-184.

Sahota, G. (1968): "Efficiency of Resource Allocation in Indian Agriculture", *American Journal of Agricultural Economics*, 50; pp. 584-605.

Saikia, Anuva (2001): "Performance of Agricultural Economy of the Northeastern India: Constraints And Priorities", in Bhuban C. Barah (ed.): Prioritization of Strategies for Agricultural Development in Northeastern India, National Centre For Agricultural Economics And Policy Research (ICAR), New Delhi.

Samuelson, Paul A. (1972): Collected Scientific Papers, MIT Press.

Schimmelpfenning, D. and C. Thirtle (1994): "Co-integration and Causalty: Exploring the Relationship between Agricultural R&D and Productivity", *Journal of Agricultural Economics*, 45(2); pp. 220-231.

Schultz, T.W. (1964): "Transforming Traditional Agriculture", Yale University Press, New Haven.

Seiford, L. and R. Thrall (1990): "Recent Development in DEA: the Mathematical Approach to Frontier Analysis", *Journal of Econometrics*, 46; pp. 7-38.

Sen, Abhijit (1992): "Economic Liberalisation and Agriculture in India", *Social Scientist*, 20(11), November; pp. 4-19.

Sen, Abhijit (2004): "Changes in Cost Structure and Farm Business Incomes, 1981-82 to 1999-00", Paper Presented in the All-India Conference on Agriculture and Rural Society in Contemporary India, Bardhman, December 17-20.

Sen, Abhijit and Jayati Ghosh (1993): "Trends in Rural Employment and the Poverty Employment Linkage", ILO ARTEP Working Papers.

Shah T. and S. Verma (2008): "Co-Management of Electricity and Groundwater: An Assessment of Gujarat's Jyoti-Gram Scheme", *Economic and Political Weekly*, 43(7); pp. 59-66.

Shah, Tushar (1993): "Groundwater Markets and Irrigation Development: Political Economy and Practical Policy", Oxford University Press: Bombay, India.

Shah, Tushar (2009): "Taming the Anarchy? Groundwater Governance in South Asia", Washington D.C.: RFF Press.

Shaik, Salim and Richard K. Perrin (1998): "Non-parametric Environmental Productivity [EAP] Measures: Nebraska Agriculture Sector", Selected Paper, American Agricultural Economics Association Meetings, Salt Lake City, Utah, August 2-5.

Shaikh, Anwar (1974): "Laws of Production and Laws of Algebra: The Humbug Production Function", *The Review of Economics and Statistics*, 56(1), pp. 115-120.

Shaikh, Anwar (1980): "Laws of Production and Laws of Algebra: Humbug II", in E.J. Nell (ed.), Growth, Profits and Property, Cambridge University Press, Cambridge.

Shaikh, Anwar (2005): "Non-linear Dynamics and Pseudo-Production Functions", *Eastern Economic Journal*, 31(5); pp. 447-466.

Shankar, U. (1998): "Econometric Application in Agricultural Economics", in K. L. Krishna (ed.): Econometric Applications in India, Oxford University Press, Delhi; pp. 1-52.

Shapiro, K. (1983): "Efficiency Differentials in Peasant Agriculture and their Implications for Development Policies", *Journal of Development Studies*, 19; pp. 179-190.

Sharma, Vijay Paul (2012): "Accelerating Agricultural Development for Inclusive Growth: Strategic Issues and Policy Options", *Vikalpa*,

January-March: 37(1); pp. 1-17.

Shephard, R.W. (1970): "Theory of Cost and Production Functions", Princeton University Press, Princeton, New Jersey.

Shergill, H.S. (2007): "Sustainability of Wheat-Rice Production in Punjab: A Re-Examination", *Economic and Political Weekly*, XLII (52), December 29; pp. 81-85.

Shiva, V. (1991): "The Green Revolution in the Punjab", *The Ecologist*, 21(2); pp. 57-60.

Shiva, Vandana (1991): "The Violence of the Green Revolution: Third World Agriculture, Ecology and Politics", London, Zed Books.

Siddiqui, Iqtidar Husain (1986): "Water Works and Irrigation System in India during Pre-Mughal Times", *Journal of the Economic and Social History of the Orient*, 29(1); pp. 52-77.

Sidhu, D.S. and Derek Byerlee (1991): "Technical Change and Wheat Productivity in Post-Green Revolution Punjab", *Economic and Political Weekly*, December 28; pp. A159-A166.

Sidhu, D. S. and Derek Byerlee (1992): "Technical Change and Wheat Productivity in the Indian Punjab in the Post-Green Revolution Period", CIMMYT Economics Working Paper 92-02, Mexico, D.F.: CIMMYT.

Sidhu, R.S., K. Vatta and H.S. Dhaliwal (2010): "Conservation Agriculture in Punjab: Economic Implications of Technologies and Practices", *Indian Journal of Agricultural Economics*, 65(3); pp. 413-427.

Sidhu, S.S. and C.A. Baanante (1979): "Farm-level Fertilizer Demand for Mexican Wheat Varieties in the Indian Punjab", *American Journal of Agricultural Economics*, 61(3), pp. 455-462.

Siebert S., J. Burke, J.M. Faures, K. Frenken, J. Hoogeveen, P. Doell and F.T. Portman (2010): "Groundwater Use For Irrigation: A Global Inventory", *Hydrology and Earth System Science,* 14; pp. 1863-1880.

Singh, R.P., A.K. Vasisht and V.C. Mathur (2003): "Quantitative Assessment of Economic Losses of Degraded Land in India", *Agricultural Economics Research Association,* India, New Delhi.

Singh, Alka and Suresh Pal (2010): "The Changing Pattern and Sources of Agricultural Growth in India", in Julian Alston, Bruce Babcock and Philip Pardey (eds.), The Shifting Patterns of Agricultural Production and Productivity Worldwide, Midwest Agribusiness Trade and Research Information Centre (MATRIC): Iowa State University, Ames, Iowa.

Singh, Amerika and O.P. Sharma (2004): "Integrated Pest Management for Sustainable Agriculture", in Pratap S. Birthal and O.P. Sharma (eds.), Proceedings, Integrated Pest Management in Indian Agriculture, NCAP, New Delhi, India.

Singh, Gurnel, Ram Babu, L.S. Bhushan and I.P. Abrol (1992): "Soil Erosion Rates in India", *Journal of Soil and Water Conservation,* 47 (1); pp. 97-99.

Singh, I.P., B. Singh and H.S. Bal (1987): "Indiscriminate Fertiliser Uses vis-à-vis Groundwater Pollution in Central Punjab", *Indian Journal of Agricultural Economics*, 42, pp. 404-409.

Singh, J. and J.P. Singh (1995): "Land Degradation and Economic Sustainability", *Ecological Economics*, 15(1); pp. 77-86.

Singh, J., P. Hone and K.K. Kundu (2002): "Measuring Inefficiencies in Crop Production on Degraded Lands", *Indian Journal of Agricultural Economics*, 57(1); January-March, pp. 65-76.

Singh, Karam and Sajla Kalra (2007): "Rice Production in Punjab: Systems, Varietal Diversity, Growth and Sustainability", *Economic and Political Weekly,* July, pp. 3139-3148.

Smale, Melinda (1997): "The Green Revolution and Wheat Genetic Diversity: Some Unfounded Assumptions", *World Development*, 25(8); pp. 1257-1269.

Solow, Robert M. (1957): "Technical Change and the Aggregate Production Function", *Review of Economics and Statistics*, August, pp. 312-320.

Staatz, J.M. and C.K. Eicher (1998): "Agricultural Development: Ideas in Historical Perspective", in Eicher, C.K. and J.M. Staatz (eds.), *International Agricultural Development,* 3rd Edition, Baltimore: Johns Hopkins University Press; pp. 8-38.

Stigler, G.J. (1976): "The Existence of X-Efficiency", *American Economic Review*, 66; pp. 213-216.

Stiglitz, J.E. (1987): "Some Theoretical Aspects of Agricultural Policies", *The World Bank Research Observer*, 2; pp. 43-60.

Syverson, Chad (2011): "What Determines Productivity?", *Journal of Economic Literature*, 49(2), pp. 326-365.

Tan, Z.X., R. Lal and K.D. Wiebe (2004): "Global Soil Nutrient Depletion and Yield Reduction", *Journal of Sustainable Agriculture*, 26(1), pp. 123-146.

Tandon, H.L.S. (2004): "Fertilizers in Indian Agriculture: From 20th to 21st Century, FDCO, New Delhi, India; p. 240.

Thanassoulis, E., M. Portela and O. Despic (2008): "Data Envelopment

Analysis: The Mathematical Programming Approach To Efficiency Analysis", in H. Fried, K. Lovell, and S. Schmidt (eds.), The Measurement of Productive Efficiency and Productivity Growth, Chapter 3; pp. 251-20; Oxford University Press.

Thiam, A., B.E. Bravo-Ureta, and T.E. Rivas (2001): "Technical Efficiency in Developing Country Agriculture: A Meta-Analysis", *Agricultural Economics*, 25; pp. 235-245.

Tiffin, R. and X. Irz (2006): "Is Agriculture the Engine of Growth?", *Agricultural Economics,* 35; pp. 79-89.

Tilman, D., J. Fargione, B. Wolff, C. DÁntonoio, A. Dobson, R. Howarth, D. Schindler, W. Schlesinger, D. Simberloff and D. Swackhamer (2001): "Forecasting Agriculturally Driven Global Environment Change", *Science*, April, 292, pp. 281-284.

Timmer, C.P. (1971): "Using a Probabilistic Frontier to Measure Technical Efficiency", *Journal of Political Economy*, 79; pp. 776-794.

Timmer, C. P. (2007): "Agricultural Growth", in Clark, D. (eds.), "The Elgar Companion to Development Studies", Edward Elgar Publishing.

Timmer, C.P. (2009): "A World without Agriculture: The Structural Transformation in Historical Perspective", Washington D.C.: AEI Press.

Timsina, J. and D.J. Connor (2001): "Productivity and Management of Rice-Wheat Cropping Systems: Issues and Challenges", *Field Crops Research*, 69; pp. 93-132.

Tobin, James (1958): "Estimation of Relationships for Limited Dependent Variables", *Econometrica*, 26(1); pp. 24-36.

Singh, J. and M. Hossain (2002): "Total Factor Productivity Analysis and its Components in a High-Potential Rice-Wheat System: A Case Study of the Indian Punjab", in M. Sombilla, M. Hossain and B. Hardy (eds.), Developments in the Asian Rice Economy, The International Rice Research Institute (IRRI), pp. 409-418.

Tripathi, A and Prasad, A.R. (2008): "Agricultural Productivity Growth in India", *Journal of Global Economy*, 4(4); pp. 322-328.

Tsakok, I. and B. Gardner (2007): "Agriculture in Economic Development: Primary Engine of Growth or Chicken and Egg?", *American Journal of Agricultural Economics*, 89(5); pp. 1145-1151.

Uphoff, N. and A. Kassam (2009): "Case Study: The System of Rice Intensification", (IP/A/STOA/FWC/2005-28/SC42), Annex 3 in

European Parliament: Science and Technology Options Assessment (STOA), June.

Vaidyanathan, A. (1994): "Performance of Indian Agriculture since Independence", in Kaushik Basu (ed.), Agrarian Questions, Oxford University Press.

Vaidyanathan, A. (1996): "Depletion of Groundwater: Some Issues", *Indian Journal of Agricultural Economics*, 51(1-2); pp. 184-192.

Vaidyanathan, A. (2000): "India's Agricultural Development Policy", *Economic and Political Weekly*, 13th May; pp. 1735-1741.

Vaidyanathan, A. (2010): "Agricultural Growth in India: The Role of Technology, Incentives and Institutions", Oxford University Press, New Delhi, India.

Varian, Hail R. (1984): "The Non-parametric Approach to the Production Analysis", *Econometrica*, 52; pp. 579-597.

Varian, Hall R. (1993): "Microeconomic Analysis", Third Edition, W. W. Norton and Company, New York.

Voelcker, John Augustus (1893): "Report on the Improvement of Indian Agriculture", Eyre & Spottiswoode, London; Reprinted by Agricole Reprints Corporation (1986), Delhi.

Wakeford, J. (2004): "The Productivity-Wage Relationship in South Africa: An empirical Investigation", *Development South Africa*, 21; pp. 109-132.

Wakerlof, G.A. (1982): "Labour Contracts as Partial Gift Exchange", *Quarterly Journal of Economics*, 97; pp. 543-569.

World Bank (1995): "Monitoring Environmental Progress: A Report on Work in Progress", Environmentally Sustainable Development Series, Washington: The World Bank.

Yadav, J.S.P. (2002): "Agricultural Resource Management in India: The Challenges", *Journal of Agricultural Resource Management*, 1; pp. 61-69.

Yadav, R.L., B.S. Dwivedi, and P.S. Pandey (2000): "Rice-Wheat Cropping System: Assessment of Sustainability under Green Manuring and Chemical Fertilizer Inputs", *Field Crops Research*, 65; pp. 15-30.

Yamada, Sahuro and Vernon W. Ruttan (1980): "International Comparison of Productivity in Agriculture", in J. Kendrick and N. Vaccara (eds.), New Development in Productivity Measurements and Analysis, Chicago University Press, Chicago.

Young, Alwyn (1992): "A Tale of Two Cities: Factor Accumulation and Technical Change in Hong Kong and Singapore", *NBER*

Macroeconomic Annual, Cambridge: MIT Press.

Young, Alwyn (1995): "The Tyranny of Numbers: Confronting the Statistical Realities of the East Asian Growth Experience", *Quarterly Journal of Economics,* 110, pp. 641-680.

Zellner, A. (1971): "An Introduction to Bayesian Inference in Econometrics", New York, John Wiley.

Zellner, A. and N. Revankar (1969): "Generalized Production Functions", *Review of Economic Studies,* 36, pp. 241-250.

Zellner, A., J. Kmenta, and J. Dreze (1966): "Specification and Estimation of Cobb-Douglas Production Function Models", *Econometrica,* 34(4); pp. 784-795.

Index

Index

A
Agricultural Demand for Water, 110
Agricultural Inputs, 42
Agricultural Productivity, 6
Agriculture Technology Management Agencies (ATMAs), 142
Agriculture-Environment Interface, 73

C
Causes of Low Productivity, 18
Central Ground Water Board, 98
Central Pollution Control Board, 131
Central Water Commission, 98, 131
Chandra Sekhar Azad, 116
Climate Change and Agriculture, 73
Cobb-Douglas Production Frontier, 60
Cropping Pattern in Indian Agriculture, 16

D
David Ricardo, 1
Department of Animal Husbandry, Dairying and Fisheries, 138
Derek Byerlee, 3
Disadvantages of Small Holdings, 15
Domestic Demand for Water, 110

E
Eighth Five Year Plan (1992-1997), 102
Eleventh Five Year Plan (2007-12), 23
Environmental Pollution, 70

F
Farm Size and Productivity, 15
Fibre Crops, 17
First Five Year Plan (1951-56), 12
Fly Ash Utilization, 103

G
Geological Survey of India (GSI), 76
Green House Gases (GHGs), 74
Green Revolution Period, 34
Green Revolution, 47
Gross Domestic Product (GDP), 22
Ground Water Irrigation, 111

H
Horticultural Crops, 18

I
Indian Institute of Soil Science (IISS), 93
Industrial Demand for

Water, 110
Integrated Watershed Management Programme, 143
Intergovernmental Panel on Climate Change (IPCC), 74
International Soil Reference and Information Centre (ISRIC), 73

L

Labour-intensive Agriculture, 39
Land Pollution and Land Degradation Causes, 82
Low-Gangetic Plain (LGP) Region, 58

M

M.S. Swaminathan, 124
Mahatma Gandhi National Rural Employment Guarantee Act (MGNREGA), 122
Malmquist Index, 66
Minimum Support Prices (MSPs), 124
Ministry of Agriculture, 99
Ministry of Environment and Forests (MoEF), 104
Ministry of Water Resources, 131

N

National Agriculture Policy (NAP), 46
National Bureau of Soil Survey, 82
National Fisheries Development Board, 138
National Food Security Mission (NFSM), 37
National Horticulture Mission, 18
National Mission for Sustainable Agriculture (NMSA), 77, 103
National Project on Management of Soil Health and Fertility (NPMSH&F), 139
National Rural Drinking Water Programme, 143
National Sample Survey Organisation (NSSO), 135
National Social Assistance Programme, 143
National Water Mission (NWM), 123

P

Pest Problems, 96
Plantation Crops, 17
Post-harvest Losses, 26
Pradhan Mantri Gram Sadak Yojana, 143
Prime Minister's Council on Climate Change, 78

R

Rajiv Gandhi Grameen Vidyutikaran Yojana, 143
Rashtriya Krishi Vikas Yojana (RKVY), 37
Reclamation and Development of Alkali and Acid Soils

(RADAS), 102
Repair, Renovation and Restoration (RRR) of Water Bodies, 122
Revamped Public Distribution System, 29
Robert Chambers, 47
Royal Commission on Agriculture, 30
Royal Commission on Irrigation, 30
Rural Infrastructural Development Fund, 29

S

Soil and Land Use Survey of India (SLUSI), 100
Soil Erosion and Productivity Loss, 88
Strategy for Efficient Management of Water Resources, 122
Sustainable Development, 72

T

T.R. Malthus, 1
T.V. Jayan, 107
Targeted Public Distribution System, 29
Theodore W. Schultz, 4
Törnqvist-Theil index, 52
Total Sanitation Campaign, 143
Twelfth Five Year Plan (2012-17), 28, 122

U

United Nations Environment Programme (UNEP), 74
Utilization of Water, 109

W

World Bank, 3
World Development Report, 3
World Meteorological Organization (WMO), 74